Scuola Normale Superiore

CATTEDRA GALILEIANA

Darrell Duffie

Credit Risk Modeling
with Affine Processes

PISA 2004

Darrell Duffie
Graduate School of Business
Stanford University
518 Memorial Way
Stanford, CA 94305-5015
USA

Credit Risk Modeling with Affine Processes

Contents

Credit Risk Modeling with Affine Processes

1. – Introduction

This is a written version of the Cattedra Galileana lectures, presented in 2002 at the Scuola Normale in Pisa. The objective is to combine an orientation to credit-risk modeling (emphasizing the valuation of corporate debt and credit derivatives) with an introduction to the analytical tractability and richness of affine state processes. This is not a general survey of either topic, but rather is designed to introduce researchers with some background in mathematics to a useful set of modeling techniques and an interesting set of applications.

Appendix A contains a brief overview of structural credit risk models, based on default caused by an insufficiency of assets relative to liabilities, including the classic Black-Scholes-Merton model of corporate debt pricing as well as a standard structural model, proposed by Fisher, Heinkel, and Zechner [1989] and solved by Leland [1994], for which default occurs when the issuer's assets reach a level so small that the issuer finds it optimal to declare bankruptcy. The alternative, and our main objective, is to treat default by a "reduced-form" approach, that is, at an exogenously specified intensity process. As a special tractable case, we often suppose that the default intensity and interest rate processes are linear with respect to an "affine" Markov state process.

Section 2 begins with the notion of default intensity, and the related calculation of survival probabilities in doubly-stochastic settings. The underlying mathematical foundations are found in Appendix E. Section 3 introduces the notion of affine processes, the main source of example calculations for the re-

This is the written version of the Cattedra Galileana lectures, Scuola Normale Superiore, in Pisa, 2002, made possible through the wonderful organizational work of Maurizio Pratelli, to whom I am most grateful. I am also grateful for support for the course offered by the Associazione Amici della Scuola Normale Superiore, who were generously represented by Mr. Carlo Gulminelli.

mainder. Technical foundations for affine processes are found in Appendix C. Section 4 explains the notion of risk-neutral probabilities, and provides the change of probability measure associated with a given change of default intensity (a version of Girsanov's theorem). Technical details for this are found in Appendix E.

By Section 5, we see the basic model for pricing defaultable debt in a setting with stochastic interest rates and stochastic risk-neutral default intensities, but assuming no recovery at default. The following section extends the pricing models to handle default recovery under alternative parameterizations. Section 7 introduces multi-entity default modeling with correlation. Section 8 turns to applications such as default swaps, credit guarantees, irrevocable lines of credit, and ratings-based step-up bonds. Appendix F provides some directions for further reading.

2. – Intensity-Based Modeling of Default

This section introduces a model for a default time as a stopping time τ with a given intensity process, as defined below. From the joint behavior of the default time, interest-rates, the promised payment of the security, and the model of recovery at default, as well as risk premia, one can characterize the stochastic behavior of the term structure of yields on defaultable bonds.

In applications, default intensities may be allowed to depend on observable variables that are linked with the likelihood of default, such as debt-to-equity ratios, volatility measures, other accounting measures of indebtedness, market equity prices, bond yield spreads, industry performance measures, and macroeconomic variables related to the business cycle, as in Duffie and Wang [2003]. This dependence could, but in practice does not usually, arise endogenously from a model of the ability or incentives of the firm to make payments on its debt. Because the approach presented here does not depend on the specific setting of a firm, it has also been applied to the valuation of defaultable sovereign debt, as in Duffie, Pedersen, and Singleton [2003] and Pagès [2000]. (For more on sovereign debt valuation, see Gibson and Sundaresan [1999] and Merrick [1999].)

We fix a complete probability space $(\Omega, \mathcal{F}, \mathbb{P})$ and a filtration $\{\mathcal{G}_t : t \geq 0\}$ of sub-σ-algebras of $\mathcal{F}$ satisfying the usual conditions, which are listed in Appendix B. Appendix E defines a nonexplosive counting process. Such a counting process K records by time t the number K_t of occurences of events of concern. Appendix E also defines the notion of a predictable process, which is, intuitively speaking, a process whose value at any time t depends only on the information in the underlying filtration $\{\mathcal{G}_t : t \geq 0\}$ that is available up to, but not including, time t.

A counting process K has an intensity λ if λ is a predictable non-negative process satisfying $\int_0^t \lambda_s \, ds < \infty$ almost surely for all t, with the property that a local martingale M, the compensated counting process, is given by

$$(2.1) \qquad M_t = K_t - \int_0^t \lambda_s \, ds.$$

Details are found in Appendix E. The accompanying intuition is that, at any time t, the $\mathcal{G}_t$-conditional probability of an event between t and $t + \Delta$ is approximately $\lambda_t \Delta$, for small Δ. This intuition is justified in the sense of derivatives if λ is bounded and continuous, and under weaker conditions.

A counting process with a deterministic intensity process is a Poisson process. If the intensity of a Poisson process is some constant α, then the times between events are independent exponentially distributed times with mean $1/\alpha$. A standard reference on counting processes is Brémaud [1981]. Additional sources include Daley and Vere-Jones [1988] and Karr [1991].

We will say that a stopping time τ has an intensity λ if τ is the first event time of a nonexplosive counting process whose intensity process is λ.

A stopping time τ is nontrivial if $\mathbb{P}(\tau \in (0, \infty)) > 0$. If a stopping time τ is nontrivial and if the filtration $\{\mathcal{G}_t : t \geq 0\}$ is the standard filtration of some Brownian motion B in $\mathbb{R}^d$, then τ could not have an intensity. We know this from the fact that if $\{\mathcal{G}_t : t \geq 0\}$ is the standard filtration of B, then the associated compensated counting process M of (2.1) (indeed, any local martingale) could be represented as a stochastic integral with respect to B, and therefore cannot jump, but M must jump at τ. In order to have an intensity, a stopping time must be totally inaccessible, a property whose definition (for example, in Meyer [1966]) suggests arrival as a "sudden surprise", but there are no such surprises on a Brownian filtration!

As an illustration, we could imagine that the equityholders or managers of a firm are equipped with some Brownian filtration for purposes of determining their optimal default time τ, as in Appendix A, but that bondholders have imperfect monitoring, and may view τ as having an intensity with respect to the bondholders' own filtration $\{\mathcal{G}_t : t \geq 0\}$, which contains less information than the Brownian filtration. Duffie and Lando [2001] provide, under conditions, the associated default intensity.[1]

We say that a stopping time τ is doubly stochastic with intensity λ if the underlying counting process whose first jump time is τ is doubly stochastic with intensity λ, as defined in Appendix E. The doubly-stochastic property implies that, for any time t, on the event that the default time τ is after t, the probability of survival to a given future time s is

$$(2.2) \qquad P(\tau > s \mid \mathcal{G}_t) = E\left[e^{- \int_t^s \lambda(u) \, du} \mid \mathcal{G}_t \right].$$

Property (2.2) is convenient for calculations, because evaluating the expectation in (2.2) is computationally equivalent to the standard financial calculation of

---

[1]Elliott, Jeanblanc, and Yor [2000] give a new proof of this intensity result, which is generalized by Song [1998] to the multi-dimensional case. Kusuoka [1999] provides an example of this intensity result that is based on unobservable drift of assets.

default-free zero-coupon bond price, treating λ as a short-term interest-rate process. Indeed, this analogy is also quite helpful for intuition when extending (2.2) to pricing applications.

It is sufficient for the convenient survival-time formula (2.2) that $\lambda_t = \Lambda(X_t)$ for some measurable $\Lambda : \mathbb{R}^d \to [0, \infty)$, where X in $\mathbb{R}^d$ solves a stochastic differential equation of the form

$$(2.3) \qquad\qquad dX_t = \mu(X_t)\,dt + \sigma(X_t)\,dB_t,$$

for some $(\mathcal{G}_t)$-standard Brownian motion B in $\mathbb{R}^d$. Here, $\mu(\cdot)$ and $\sigma(\cdot)$ are functions on the state space of X that satisfy enough regularity for (2.3) to have a unique (strong) solution. With this, the survival probability calculation (2.2) is of the form

$$(2.4) \qquad \mathbb{P}(\tau > s \mid \mathcal{G}_t) = E\left[e^{-\int_t^s \Lambda(X(u))\,du} \mid X(t)\right]$$
$$(2.5) \qquad\qquad\qquad\qquad = f(X(t), t),$$

where, under the usual regularity for the Feynman-Kac approach, $f(\cdot)$ solves the partial differential equation (PDE)

$$(2.6) \qquad\qquad \mathcal{A}f(x, t) - f_t(x, t) - \Lambda(x)f(x, t) = 0,$$

for the generator $\mathcal{A}$ of X, given by

$$\mathcal{A}f(x, t) = \sum_i \frac{\partial}{\partial x_i} f(x, t)\mu_i(x) + \frac{1}{2} \sum_{i,j} \frac{\partial^2}{\partial x_i\,\partial x_j} f(x, t)\gamma_{ij}(x),$$

and where $\gamma(x) = \sigma(x)\sigma(x)'$, with the boundary condition

$$(2.7) \qquad\qquad\qquad f(x, s) = 1.$$

Parametric assumptions are often used to get an explicit solution to this PDE, as we shall see.

More generally, (2.2) follows from assuming that the doubly-stochastic counting process K whose first jump time is τ is driven by some filtration $\{\mathcal{F}_t : t \geq 0\}$, a concept defined in Appendix E. (Included in the definition is the condition that $\mathcal{F}_t \subset \mathcal{G}_t$, and that $\{\mathcal{F}_t : t \geq 0\}$ satisfies the usual conditions.) The intuition of the doubly-stochastic assumption is that $\mathcal{F}_t$ contains enough information to reveal the intensity λ_t, but not enough information to reveal the event times of the counting process K. In particular, at any current time t and for any future time s, after conditioning on the σ-algebra $\mathcal{G}_t \vee \mathcal{F}_s$ generated by the events in $\mathcal{G}_t \cup \mathcal{F}_s$, K is a Poisson process up to time s with (conditionally deterministic) time-varying intensity $\{\lambda_t : 0 \leq t \leq s\}$, so the number $K_s - K_t$ of arrivals between t and s is therefore conditionally distributed as a Poisson random variable with parameter $\int_t^s \lambda_u\,du$. (A random variable q has the Poisson distribution with parameter β if $\mathbb{P}(q = k) = e^{-\beta}\beta^k/k!$ for any nonnegative integer k.) Thus, letting A be the event $\{K_s - K_t = 0\}$ of no arrivals, the law of iterated expectations implies that, for $t < \tau$,

$$
\begin{aligned}
\mathbb{P}(\tau > s \mid \mathcal{G}_t) &= E(1_A \mid \mathcal{G}_t) \\
&= E[E(1_A \mid \mathcal{G}_t \vee \mathcal{F}_s) \mid \mathcal{G}_t] \\
&= E\left[\mathbb{P}(K_s - K_t = 0 \mid \mathcal{G}_t \vee \mathcal{F}_s) \mid \mathcal{G}_t\right] \\
&= E\left[e^{\int_t^s -\lambda(u)\,du} \mid \mathcal{G}_t\right],
\end{aligned}
$$

(2.8)

consistent with (2.2). Appendix E connects the intensity of τ with its probability density function and its hazard rate.

3. – Affine Processes

In many financial applications that are based on a state process, such as the solution X of (2.3), a useful assumption is that the state process X is "affine." An affine process X with some state space $D \subset \mathbb{R}^d$ is a Markov process whose conditional characteristic function is of the form, for any $u \in \mathbb{R}^d$,

$$
(3.1) \qquad E\left(e^{iu \cdot X(t)} \mid X(s)\right) = e^{\varphi(t-s,iu) + \psi(t-s,iu) \cdot X(s)},
$$

for some coefficients $\varphi(\cdot, iu)$ and $\psi(\cdot, iu)$. We will take the state space D to be of the standard form $\mathbb{R}_+^n \times \mathbb{R}^{d-n}$, for $0 \le n \le d$. We say that X is "regular" if the coefficients $\varphi(\cdot, iu)$ and $\psi(\cdot, iu)$ of the characteristic function are differentiable and if their derivatives are continuous at 0. This regularity implies that these coefficients satisfy a generalized Riccati ordinary differential equation (ODE) given in Appendix C. The form of this ODE in turn implies, roughly speaking, that X must be a jump-diffusion process, in that

$$
(3.2) \qquad dX_t = \mu(X_t)\,dt + \sigma(X_t)\,dB_t + dJ_t,
$$

for a standard Brownian motion B in $\mathbb{R}^d$ and a pure-jump process J, such that the drift $\mu(X_t)$, the "instantaneous" covariance matrix $\sigma(X_t)\sigma(X_t)'$, and the jump measure associated with J all have affine dependence on the state X_t. Conversely, jump-diffusions of this form (3.2) are affine processes in the sense of (3.1). A more careful statement of this result is found in Appendix C.

Simple examples of affine processes used in financial modeling are the Gaussian Ornstein-Uhlenbeck model, applied to interest rates by Vasicek [1977], and the Feller [1951] diffusion, applied to interest-rate modeling by Cox, Ingersoll, and Ross [1985]. A general multivariate class of affine jump-diffusion models was introduced by Duffie and Kan [1996] for term-structure modeling. Using 3-dimensional affine diffusion models, for example, Dai and Singleton [2000] found that both time-varying conditional variances and negatively correlated state variables are essential ingredients to explaining the historical behavior of term structures of U.S. interest rates.

For option pricing, there is a substantial literature building on the particular affine stochastic-volatility model for currency and equity prices proposed by Heston [1993]. Bates [1997], Bakshi, Cao, and Chen [1997], Bakshi and Madan [2000], and Duffie, Pan, and Singleton [2000] brought more general affine models to bear in order to allow for stochastic volatility and jumps, while maintaining and exploiting the simple property (3.1).

A key property related to (3.1) is that, for any affine function $\Lambda : D \to \mathbb{R}$ and any $w \in \mathbb{R}^d$, subject only to technical conditions reviewed in Duffie, Filipović, and Schachermayer [2003],

$$(3.3) \qquad E_t \left[e^{\int_t^s -\Lambda(X(u))\, du + w \cdot X(s)} \right] = e^{\alpha(s-t)+\beta(s-t)\cdot X(t)},$$

for coefficients $\alpha(\cdot)$ and $\beta(\cdot)$ that satisfy generalized Riccati ODEs (with real boundary conditions) of the same type solved by φ and ψ of (3.1), respectively.

In order to get a quick sense of how (3.3) and the associated Riccati equations for the solution coefficients $\alpha(\cdot)$ and $\beta(\cdot)$ arise, we consider the special case of an affine diffusion process X solving the stochastic differential equation (2.3), with state space $D = \mathbb{R}_+$, and with $\mu(x) = a + bx$ and $\sigma^2(x) = cx$, for constant coefficients a, b, and c. (This is the continuous branching process of Feller [1951].) We let $\Lambda(x) = \rho_0 + \rho_1 x$, for constants ρ_0 and ρ_1, and apply the (Feynman-Kac) PDE (2.6) to the candidate solution (3.3). After calculating all terms of the PDE and then dividing each term of the PDE by the common factor $f(x, t)$, we arrive at

$$(3.4) \qquad -\alpha'(z) - \beta'(z)x + \beta(z)(a + bx) + \frac{1}{2}\beta(z)^2 c^2 x - \rho_0 - \rho_1 x = 0,$$

for all $z \geq 0$. Collecting terms in x, we have

$$(3.5) \qquad u(z)x + v(z) = 0,$$

where

$$(3.6) \qquad u(z) = -\beta'(z) + \beta(z)b + \frac{1}{2}\beta(z)^2 c^2 - \rho_1$$

$$(3.7) \qquad v(z) = -\alpha'(z) + \beta(z)a - \rho_0.$$

Because (3.5) must hold for all x, it must be the case that $u(z) = v(z) = 0$. (This is known as "separation of variables.") This leaves the Riccati equations:

$$(3.8) \qquad \beta'(z) = \beta(z)b + \frac{1}{2}\beta(z)^2 c^2 - \rho_1$$

$$(3.9) \qquad \alpha'(z) = \beta(z)a - \rho_0,$$

with the boundary conditions $\alpha(0) = 0$ and $\beta(0) = w$, from the boundary condition $f(x, s) = w$ for all x. The explicit solutions for $\alpha(z)$ and $\beta(z)$, developed by Cox, Ingersoll, and Ross [1985] for bond pricing (that is, for $w = 0$), is repeated in Appendix D, in the context a slightly more general model with jumps.

The calculation (3.3) arises in many financial applications, some of which will be reviewed momentarily. An obvious example is discounted expected cash flow (with discount rate $\Lambda(X_t)$), as well as the survival-probability calcuation (2.2) for an affine state process X and a default intensity $\Lambda(X_t)$, taking $w = 0$ in (3.3).

3.1. – Examples of Affine Processes

An affine diffusion is a solution X of the stochastic differential equation of the form (2.3) for which both $\mu(x)$ and $\sigma(x)\sigma(x)'$ are affine in x. This class includes the Gaussian (Ornstein-Uhlenbeck) case, for which $\sigma(x)$ is constant (used by Vasicek [1977] to model interest rates), as well as the Feller [1951] diffusion model, used by Cox, Ingersoll, and Ross [1985] to model interest rates. These two examples are one-dimensional; that is, $d = 1$. For the case in which X is a Feller diffusion, we can write

$$(3.10) \qquad dX_t = \kappa(\bar{x} - X_t)\,dt + c\sqrt{X_t}\,dB_t,$$

constant positive parameters[2] c, κ, and $\bar{x}$. The parameter $\bar{x}$ is called a "long-run mean," and the parameter κ is called the mean-reversion rate. Indeed for (3.10), the mean of X_t converges from any initial condition to $\bar{x}$ at the rate κ as t goes to ∞. The Feller diffusion, originally conceived as a continuous branching process in order to model randomly fluctuating population sizes, has become popularized in finance as the "Cox-Ingersoll-Ross" (CIR) process.

Beyond the Gaussian case, any Ornstein-Uhlenbeck process, whether driven by a Brownian motion (as for the Vasicek model) or by a more general Lévy process, as in Sato [1999], is affine. Moreover, any continuous-branching process with immigration (CBI process), including multi-type extensions of the Feller process, is affine. (See Kawazu and Watanabe [1971].) Conversely, as stated in Appendix C, an affine process in $\mathbb{R}_+^d$ is a CBI process.

A special example of (3.2) is the "basic affine process," with state space $D = \mathbb{R}_+$, satisfying

$$(3.11) \qquad dX_t = \kappa(\bar{x} - X_t)\,dt + c\sqrt{X_t}\,dB_t + dJ_t,$$

where J is a compound Poisson process,[3] independent of B, with exponential jump sizes. The Poisson arrival intensity λ of jumps and the mean γ of the jump

---

[2] The solution X of (3.10) will never reach zero from a strictly positive initial condition if $\kappa\bar{x} > c^2/2$, which is sometimes called the "Feller condition."

[3] A compound Poisson process has jumps at iid exponential event times, with iid jump sizes.

sizes completes the list $(\kappa, \bar{x}, c, \bar{\lambda}, \gamma)$ of parameters of a basic affine process. Special cases of the basic affine model include the model with no diffusion $(c = 0)$ and the diffusion of Feller [1951] (for $\bar{\lambda} = 0$). The basic affine process is especially tractable, in that the coefficients $\alpha(t)$ and $\beta(t)$ of (3.3) are known explicitly, and recorded in Appendix D.4. The coefficients $\varphi(t, iu)$ and $\psi(t, iu)$ of the characteristic function (3.1) are of the same form, albeit complex.

A simple class of multivariate affine processes is obtained by letting $X_t = (X_{1t}, \dots, X_{dt})$, for independent affine coordinate processes $X_1, \dots, X_d$. The independence assumption implies that we can break the calculation (3.3) down as a product of terms of the same form as (3.3), but for the one-dimensional coordinate processes. This is the basis of the "multi-factor CIR model," often used to model interest rates, as in Chen and Scott [1995].

An important 2-dimensional affine model was used by Heston [1993] to model option prices in settings with stochastic volatility. Here, one supposes that the underlying price process U of an asset satisfies

$$(3.12) \qquad dU_t = U_t(\gamma_0 + \gamma_1 V_t)\, dt + U_t\sqrt{V_t}\, dB_{1t},$$

where γ_0 and γ_1 are constants and V is a stochastic-volatility process, which is a Feller diffusion satisfying

$$(3.13) \qquad dV_t = \kappa(\bar{v} - V_t)\, dt + c\sqrt{V_t}\, dZ_t,$$

for constant coefficients κ, $\bar{v}$, and c, where $Z = \rho B_1 + \sqrt{1 - \rho^2}\, B_2$ is a standard Brownian motion that is constructed as a linear combination of independent standard Brownian motions B_1 and B_2. The correlation coefficient ρ generates what is known as "volatility asymmetry," and is usually measured to be negative for major market stock indices. Option implied-volatility "smile curves" are, roughly speaking, rotated clockwise into "smirks" as ρ becomes negative. Letting $Y = \log U$, a calculation based on Itô's formula (see Appendix B) yields

$$(3.14) \qquad dY_t = \left(\gamma_0 + \left(\gamma_1 - \frac{1}{2} V_t \right) \right) dt + \sqrt{V_t}\, dB_{1t},$$

which implies that the 2-dimensional process $X = (V, Y)$ is affine, with state space $D = \mathbb{R}_+ \times \mathbb{R}$. By virute of the explicit characteristic function of $\log U_t$, this leads to a simple method for pricing options, as explained in Section 8. Extensions allowing for jumps have also been useful for the statistical analysis of stock returns from time-series data on underlying asset returns and of option prices, as in Bates [1996] and Pan [2002].[4]

---

[4]Among analyses of option pricing for the case of affine state variables are Bates [1997], Bakshi, Cao, and Chen [1997], Bakshi and Madan [2000], Duffie, Pan, and Singleton [2000], and Scott [1997].

4. – Risk-Neutral Probability and Intensity

Basic to the theory of the market valuation of financial securities are "risk-neutral probabilities," artificially chosen probabilities under which the price of any security is the expectation of the discounted cash flow of the security, as will be made more precise shortly.

We will assume the existence of a short-rate process, a progressively measurable process r with the property that $\int_0^t |r(u)|\, du < \infty$ for all t, and such that, for any times s and $t > s$, an investment of one unit of account (say, one Euro) at any time s, reinvested continually in short-term lending until any time t after s, will yield a market value of $e^{\int_s^t r(u)\, du}$. When we say "under $\mathbb{Q}$," for an equivalent[5] probability measure $\mathbb{Q}$, we mean with respect to the probability space $(\Omega, \mathcal{F}, \mathbb{Q})$ and the same given filtration $\{\mathcal{G}_t : t \geq 0\}$.

For the purpose of market valuation, we fix some equivalent martingale measure $\mathbb{Q}$, based on discounting at the short rate r. This means that, as of any time t, for any stopping time T and bounded $\mathcal{G}_T$-measurable random variable F, a security paying F at T has a modeled price, on the event $\{T > t\}$, of $E_t^{\mathbb{Q}}[e^{\int_t^T -r(u)\, du} F]$, where, for convenience, we write $E_t^{\mathbb{Q}}$ for expectation under $\mathbb{Q}$, given $\mathcal{G}_t$. Uniqueness of would be unexpected in a setting of default risk. an equivalent martingale measure, Harrison and Kreps [1979] showed that the existence of an equivalent martingale measure is equivalent (up to technical conditions) to the absence of arbitrage. Delbaen and Schachermayer [1999] gave definitive technical definitions and conditions for this result. There may be more than one equivalent martingale measure, however, and for modeling purposes, one would work under one such measure. Common devices for estimating an equivalent martingale measure include statistical analysis of historical price data, or the modeling of market equilibrium. If markets are complete, meaning roughly that any contingent cash flow can be replicated by trading the available securities, the equivalent martingale measure is unique (in a certain technical sense), and can be deduced from the price processes of the available securities. For further treatment, see, for example, Duffie [2001].

A risk-neutral intensity process for a default time τ is an intensity process $\lambda^{\mathbb{Q}}$ for the default time τ, under $\mathbb{Q}$. We also call $\lambda^{\mathbb{Q}}$ the $\mathbb{Q}$-intensity of τ. Artzner and Delbaen [1995] gave us the following convenient result.

PROPOSITION. *Suppose that a nonexplosive counting process K has a $\mathbb{P}$-intensity process, and that $\mathbb{Q}$ is any probability measure equivalent to $\mathbb{P}$. Then K has a $\mathbb{Q}$-intensity process.*

The ratio $\lambda^{\mathbb{Q}}/\lambda$ (for λ strictly positive) represents a risk premium for uncertainty associated with the timing of default, in the sense of the following version of Girsanov's theorem, which provides conditions suitable for calculating the change of probability measure associated with a change of intensity, by analogy

[5] A probability measure $\mathbb{Q}$ is equivalent to $\mathbb{P}$ if $\mathbb{P}$ and $\mathbb{Q}$ assign zero probabilities to the same events in $\mathcal{G}_t$, for each t.

with the "change in drift" of a Brownian motion. Suppose K is a nonexplosive counting process with intensity λ, and that φ is a strictly positive predictable process such that, for some fixed time horizon T, $\int_0^T \varphi_s \lambda_s \, ds$ is finite almost surely. A local martingale ξ is then well defined by

$$(4.1) \qquad \xi_t = \exp\left(\int_0^t (1 - \varphi_s)\lambda_s \, ds\right) \prod_{\{i:T(i)\leq t\}} \varphi_{T(i)}, \quad t \leq T.$$

GIRSANOV'S THEOREM. *Suppose the local martingale ξ is actually a martingale. Then an equivalent probability measure $\mathbb{Q}$ is defined by $\frac{d\mathbb{Q}}{d\mathbb{P}} = \xi(T)$. Restricted to the time interval $[0, T]$, the counting process K has $\mathbb{Q}$-intensity $\lambda\varphi$.*

A proof may be found in Brémaud [1981]. Care must be taken with assumptions, for the convenient doubly-stochastic property need not be preserved with a change to an equivalent probability measure. Kusuoka [1999] gives examples of this failure. Appendix E gives sufficient conditions for the martingale property of ξ, and for K to be doubly stochastic under both $\mathbb{P}$ and $\mathbb{Q}$.

Under certain conditions on the filtration $\{\mathcal{G}_t : t \geq 0\}$ outlined in Appendix E, the martingale representation property applies, and for any equivalent probability measure $\mathbb{Q}$, one can obtain the associated $\mathbb{Q}$-intensity of K from the martingale representation of the associated density process.

5. – Zero-Recovery Bond Pricing

We consider the valuation of a security that pays $F1_{\{\tau>s\}}$ at a given time $s > 0$, where F is a $\mathcal{G}_s$-measurable bounded random variable. As $1_{\{\tau>s\}}$ is the random variable that is 1 in the event of no default by s and zero otherwise, we may view F as the contractually promised payment of a security with the property that, in the event of default before the contractual maturity date s, there is no payment (that is, zero default recovery). The case of a defaultable zero-coupon bond is treated by letting $F = 1$. In the next section, we will consider non-zero recovery at default.

From the definition of $\mathbb{Q}$ as an equivalent martingale measure, the price S_t of this security at any time $t < s$ is given by

$$(5.1) \qquad S_t = E_t^{\mathbb{Q}}\left[e^{-\int_t^s r(u)\,du} 1_{\{\tau>s\}} F\right].$$

From (5.1) and the fact that τ is a stopping time, S_t must be zero for all $t \geq \tau$. The following result is based on Lando [1994]. (See, also, Duffie, Schroder, and Skiadas [1996] and Lando [1998].)

THEOREM 1. *Suppose that F, r, and $\lambda^{\mathbb{Q}}$ are bounded and that, under $\mathbb{Q}$, τ is doubly stochastic driven by a filtration $\{\mathcal{F}_t : t \geq 0\}$, with intensity process $\lambda^{\mathbb{Q}}$.*

Suppose, moreover, that r is $(\mathcal{F}_t)$-adapted and F is $\mathcal{F}_s$-measurable. Fix any $t < s$. Then, for $t \geq \tau$, we have $S_t = 0$, and for $t < \tau$,

$$(5.2) \qquad S_t = E_t^{\mathbb{Q}} \left[e^{-\int_t^s (r(u)+\lambda^{\mathbb{Q}}(u))\,du}\, F \right].$$

The idea of (5.2) is that discounting for default that occurs at an intensity is analogous to discounting at the short rate r.

PROOF. From (5.1), the law of iterated expectations, and the assumption that r is $(\mathcal{F}_t)$-adapted and F is $\mathcal{F}_s$-measurable,

$$S_t = E^{\mathbb{Q}}(E^{\mathbb{Q}}[e^{-\int_t^s r(u)\,du}\, 1_{\{\tau>s\}} F \mid \mathcal{F}_s \vee \mathcal{G}_t] \mid \mathcal{G}_t)$$
$$= E^{\mathbb{Q}}(e^{-\int_t^s r(u)\,du}\, F E^{\mathbb{Q}}[1_{\{\tau>s\}} \mid \mathcal{F}_s \vee \mathcal{G}_t] \mid \mathcal{G}_t).$$

The result then follows from the implication of double stochasticity that, on the event $\{\tau > t\}$, we have $\mathbb{Q}(\tau > s \mid \mathcal{F}_s \vee \mathcal{G}_t) = e^{\int_t^s -\lambda^{\mathbb{Q}}(u)\,du}$.   □

As a special case, suppose the driving filtration $\{\mathcal{F}_t : t \geq 0\}$ is that generated by a process X that is affine under $\mathbb{Q}$, with state space D. It is then natural to allow dependence of $\lambda^{\mathbb{Q}}$, r, and F on the state process X in the sense that

$$(5.3) \qquad \lambda_t^{\mathbb{Q}} = \Lambda(X_{t-}), \quad r_t = \rho(X_t), \quad F = e^{u \cdot X(s)},$$

where Λ and ρ are real-valued affine functions on D, and $u \in \mathbb{R}^d$. We have already adopted the convention that an intensity process is predictable, and have therefore defined $\lambda_t^{\mathbb{Q}}$ to depend in (5.3) on the left limit X_{t-}, rather than X_t itself, because X need not itself be a predictable process. For the calculation (5.2), however, this makes no difference, because $\int \Lambda(X_{t-})\,dt = \int \Lambda(X_t)\,dt$, given that $X(\omega, t-) = X(\omega, t)$ for almost every t.

With (5.2) and (5.3), we can apply the basic property (3.3) of affine processes, so that for $t < \tau$, under mild regularity,

$$S_t = e^{\alpha(s-t)+\beta(s-t)\cdot X(t)},$$

for coefficients $\alpha(\cdot)$ and $\beta(\cdot)$ satisfying the integro-differential equation associated with (5.2), namely

$$(5.4) \qquad \frac{\mathcal{A}e^{\alpha(s-t)+\beta(s-t)\cdot x}}{e^{\alpha(s-t)+\beta(s-t)\cdot x}} - \alpha'(s-t) - \beta'(s-t) \cdot x - \rho(x) - \Lambda(x) = 0,$$

where $\mathcal{A}$ is the generator (under $\mathbb{Q}$) of X, with the boundary condition $e^{\alpha(0)+\beta(0)\cdot x} = 1$. This in turn implies, by the same separation-of-variables argument used in the simple example of Section 3, an associated generalized Riccati ODE for $\alpha(\cdot)$ and $\beta(\cdot)$, with boundary condition $\alpha(0) = 0$ and $\beta(0) = 0 \in \mathbb{R}^d$. Solution of the ODE is explicit in certain cases, and otherwise can be computed routinely, say by a Runge-Kutta method.

A sufficient regularity condition for this solution is that X is a regular affine process and that the short-rate process r is non-negative. (See Duffie, Filipović, and Schachermayer [2003] for details.)

6. – Pricing with Recovery at Default

The next step is to consider the recovery of some random payoff W at the default time τ, if default occurs before the maturity date s of the security. We adopt the assumptions of Theorem 1, and add the assumption that $W = w_\tau$, where w is a bounded predictable process that is also adapted to the driving filtration $\{\mathcal{F}_t : t \geq 0\}$.

At any time t before default, the market value of the default recovery is by the definition of the equivalent martingale measure $\mathbb{Q}$,

$$(6.1) \qquad J_t = E^{\mathbb{Q}}\big[e^{-\int_t^\tau r(u)\,du} 1_{\{\tau \leq s\}} w_\tau \mid \mathcal{G}_t\big].$$

The assumption that τ is doubly-stochastic implies that it has a probability density under $\mathbb{Q}$, at any time u in $[t, s]$, conditional on $\mathcal{G}_t \vee \mathcal{F}_s$, on the event that $\tau > t$, of

$$q(t, u) = e^{-\int_t^u \lambda^{\mathbb{Q}}(z)\,dz} \lambda_u^{\mathbb{Q}}.$$

(For details, see Appendix E.) Thus, using the same iterated-expectations argument of the proof of Theorem 1, we have, on the event that $\tau > t$,

$$\begin{aligned} J_t &= E^{\mathbb{Q}}\big(E^{\mathbb{Q}}\big[e^{-\int_t^\tau r(z)\,dz} 1_{\{\tau \leq s\}} w_\tau \mid \mathcal{F}_s \vee \mathcal{G}_t\big] \mid \mathcal{G}_t\big) \\ &= E^{\mathbb{Q}}\left(\int_t^s e^{-\int_t^u r(z)\,dz} q(t, u) w_u \, du \mid \mathcal{G}_t\right) \\ &= \int_t^s \Phi(t, u)\, du, \end{aligned}$$

using Fubini's theorem, where

$$(6.2) \qquad \Phi(t, u) = E_t^{\mathbb{Q}}\big[e^{-\int_t^u (\lambda^{\mathbb{Q}}(z) + r(z))\,dz} \lambda_u^{\mathbb{Q}} w_u\big].$$

We summarize the main defaultable valuation result as follows.

THEOREM 2. *Consider a security that pays F at s if $\tau > s$, and otherwise pays w_τ at τ. Suppose that w, F, $\lambda^{\mathbb{Q}}$, and r are bounded. Suppose that τ is doubly stochastic under $\mathbb{Q}$ driven by a filtration $\{\mathcal{F}_t : t \geq 0\}$ with the property that r and w are $(\mathcal{F}_t)$-adapted and F is $\mathcal{F}_s$-measurable. Then, for $t \geq \tau$, we have $S_t = 0$, and for $t < \tau$,*

$$(6.3) \qquad S_t = E_t^{\mathbb{Q}}\big[e^{-\int_t^s (r(u) + \lambda^{\mathbb{Q}}(u))\,du} F\big] + \int_t^s \Phi(t, u)\, du.$$

6.1. – Unpredictable Default Recovery

Schonbucher [1998] extends to the case of a default recovery W that is not of the form w_τ for some predictable process w, but rather allows the recovery to be revealed just at the default time τ. We now allow this, taking however a different construction. We let T be the stopping time $\min(\tau, s)$, and let $\hat{W} = E^{\mathbb{Q}}(W 1_{\{\tau < s\}} | \mathcal{G}_{T-})$. From[6] Dellacherie and Meyer [1978], Theorem IV.67(b), there is a $(\mathcal{G}_t)$-predictable process w satisfying $w(T) = \hat{W}$. Then, for $t < \tau$, by the law of iterated expectations, the market value of default recovery is

$$(6.4) \qquad J_t = E_t^{\mathbb{Q}} \left(e^{-\int_t^T r(u)\,du} W 1_{\{\tau < s\}} \right)$$

$$(6.5) \qquad = E_t^{\mathbb{Q}} \left(e^{-\int_t^T r(u)\,du} \hat{W} \right)$$

$$(6.6) \qquad = E_t^{\mathbb{Q}} \left(e^{-\int_t^T r(u)\,du} w(T) \right)$$

$$(6.7) \qquad = E_t^{\mathbb{Q}} \left(\int_t^s e^{-\int_t^u [r(z) + \lambda^{\mathbb{Q}}(z)]\,dz} \lambda^{\mathbb{Q}}(u) w(u)\,du \right),$$

and we are back in the setting of Theorem 2. Intuitively speaking, $w(u)$ is the risk-neutral expected recovery given the information available in the filtration $\{\mathcal{G}_t : t \geq 0\}$ up until, but not including, time u, *and* given that default will occur in "the next instant," at time u.

In the affine state-space setting described at the end of the previous section, $\Phi(t, u)$ can be computed by our usual "affine" methods, provided that $w(t)$ is of the form $e^{a(t) + b(t) \cdot X(t-)}$ for deterministic $a(t)$ and $b(t)$. (In applications, a common assumption is that $w(t)$ is deterministic, but the evidence favors significant negative correlation, on average, between default recovery rates and the average rate of default itself; see Altman, Brady, Resti, and Sironi [2003].) With recovery and default intensity correlation, it is an exercise to show that, under technical regularity,

$$(6.8) \qquad \Phi(t, u) = e^{\alpha(t,u) + \beta(t,u) \cdot X(t)} [c(t, u) + C(t, u) \cdot X(t)],$$

for readily computed deterministic coefficients α, β, c, and C. (For this "extended affine" calculation, see Duffie, Pan, and Singleton [2000].) This leaves the numerical task of computing $\int_t^s \Phi(t, u)\,du$, say by quadrature.

For the price of a typical defaultable bond promising periodic coupons followed by its principal at maturity, one may sum the prices of the coupons and of the principal, treating each of these payments as though it were a separate zero-coupon bond. An often-used assumption, although one that need not apply in practice, is that there is no default recovery for coupons, and that all bonds of the same seniority (priority in default) have the same recovery of principal, regardless of maturity. In any case, convenient parametric assumptions, based for example on an affine driving process X, lead to straightforward computation of a term structure of defaultable bond yields that may be applied in practical situations.

[6] The definition of $\mathcal{G}_{T-}$ is also given in Dellacherie and Meyer [1978]. Please note that there is a typographical error in Dellacherie and Meyer [1978], Theorem IV.67(b), in that the second sentence should read: "Conversely, if Y is an $\mathcal{F}_{T-}^0$-measurable..." rather than "Conversely, if Y is an $\mathcal{F}_T^0$-measurable...," as can be verified from the proof, or, for example, from their Remark 68(b).

6.2. – Option-Embedded Corporate Bonds

For the case of defaultable bonds with embedded American options, the most typical cases being callable bonds or convertible bonds, the usual resort is valuation by some numerical implementation of the associated dynamic programming problems for optimal exercise timing. Acharya and Carpenter [2001] and Berndt [2002] treat callable defaultable bonds. On the related problem of convertible bond valuation, see Davis and Lischka [1999], Loshak [1996], Nyborg [1996], and Tsiveriotis and Fernandes [1998]. On the empirical timing behavior of call and conversion options on convertible bonds, see Ederington, Caton and Campbell [1997]

6.3. – Default-Adjusted Short Rate

In the setting of Theorem 6, a particularly simple pricing representation can be based on the definition of a predictable process ℓ for the fractional loss in market value at default, defined by

$$(6.1) \qquad\qquad (1 - \ell_\tau)(S_{\tau-}) = w_\tau.$$

Manipulation that is left as an exercise shows that, under the conditions of Theorem 2, for t before the default time,

$$(6.2) \qquad\qquad S_t = E_t^{\mathbb{Q}}\left[e^{-\int_t^s [r(u) + \ell(u)\lambda^{\mathbb{Q}}(u)]\,du}\, F\right].$$

This valuation model (6.2) is from Duffie, Schroder, and Skiadas [1996], extending Pye [1974], Litterman and Iben [1991], and Duffie and Singleton [1999]. This is particularly convenient if we take ℓ as an exogenously given fractional loss process, as it allows for the application of standard valuation methods, treating the payoff F as default-free, but accounting for the intensity and severity of default losses through the "default-adjusted" short-rate process $r + \ell\lambda^{\mathbb{Q}}$. Naturally, the discount-rate adjustment $\ell\lambda^{\mathbb{Q}}$ is the risk-neutral mean rate of proportional loss in market value due to default. Collin-Dufresne Goldstein, and Helwege [2002] extend this result to settings in which the doubly-stochastic assumption fails, by an additional change of measure under which there are almost surely no jumps.

Notably, the dependence of the bond price on the intensity $\lambda^{\mathbb{Q}}$ and fractional loss ℓ at default is only through the product $\ell\lambda^{\mathbb{Q}}$. Thus, for any bounded strictly positive predictable process θ, bounded away from zero, the price process S of (6.2) is invariant (before default), to a substitution for ℓ and $\lambda^{\mathbb{Q}}$ of $\theta\ell$ and $\lambda^{\mathbb{Q}}/\theta$, respectively. For example, doubling $\lambda^{\mathbb{Q}}$ and halving ℓ has no effect on the price process before default.

Suppose, for example, that τ is doubly stochastic driven by X, and we take $r_t + \ell_t\lambda_t^{\mathbb{Q}} = R(X_{t-})$ and $F = f(X_s)$, for a Markov state process X. For example, X could be given as the solution to (2.3), or be an affine process.

Then, under typical Feynman-Kac regularity conditions, we obtain at each time t before default the bond price $S_t = g(X_t, t)$, for a solution g of the

$$(6.3) \qquad \mathcal{A}g(x, t) - g_t(x, t) - R(x)g(x, t) = 0, \quad (x, t) \in D \times [0, s),$$

where $\mathcal{A}$ is the generator of X, with boundary condition $g(x, s) = f(x)$.

If the driving process X is affine, if $f(x) = e^{u \cdot x}$ for some $u \in \mathbb{R}^d$, and if $R(x) = a + b \cdot x$ for some $a \in \mathbb{R}$ and b in $\mathbb{R}^d$, then we have $g(x, t) = e^{\alpha(s-t)+\beta(s-t) \cdot x}$ for $\alpha(\cdot)$ and $\beta(\cdot)$ computed from the generalized Riccati equations associated with (6.3), with boundary conditions $\alpha(0) = 0$ and $\beta(0) = u$. Sufficient regularity is that X is a regular affine process and that the default-adjusted short rate $R(x)$ is non-negative.

There are also interesting cases, for example the pricing of swaps with two-sided default risk (see Duffie and Huang [1996]), for which λ_t^Q or ℓ_t (or both) are naturally dependent on the price itself. The PDE (6.3) would then be generalized to one of the nonlinear form

$$(6.4) \qquad \mathcal{A}g(x, t) - R(x, g(x, t))g(x, t) = 0.$$

For empirical work on default-adjusted short rates, see Duffee [1999] and Bakshi, Madan, and Zhang [2001] for applications to corporate bonds, and, for applications to sovereign debt, Duffie, Pedersen, and Singleton [2003] and Pagès [2000].

7. – Correlated Default

Extending, suppose that the default times $\tau_1, \ldots, \tau_k$ of k given names have respective intensity processes $\lambda_1, \ldots, \lambda_k$, and are doubly stochastic, driven by a filtration $\{\mathcal{F}_t : t \geq 0\}$, as defined in Appendix E. This means roughly that, conditional on the information in the driving filtration that determines the respective intensities, the event times $\tau_1, \ldots, \tau_k$ are independent. In particular, the only source of correlation of the default times is via the correlation of the intensities. Das, Duffie, and Kapadia [2004] provide a statistical test of this multi-name doubly-stochastic property.

One can see from the definition of an intensity process that, under the doubly-stochastic assumption, the first default time $\tau = \min(\tau_1, \ldots, \tau_n)$ has the intensity $\lambda_1(t) + \cdots + \lambda_k(t)$. Indeed, the same result applies if we weaken the doubly stochastic assumption to merely the assumption that $\tau_i \neq \tau_j$, or more precisely that, for any i and $j \neq i$, we have $\mathbb{P}(\tau_i = \tau_j) = 0$. For an easy proof, based again on the definition of intensity, see Duffie [1998b].

More generally, the doubly stochastic assumption makes the computation of the joint distribution of default times rather simple. Consider the joint survivorship event $\{\tau_1 \geq t(1), \ldots, \tau_k \geq t(k)\}$, for deterministic times $t(1), \ldots, t(k)$. Without loss of generality after relabeling, we can suppose that $t(1) \leq t(2) \leq \cdots \leq t(k)$. (We can also allow $t(i) = +\infty$.) Then, by the doubly stochastic assumption, for any "current" time $t < t(1)$, on the event that $\tau_i > t$ for all i (no defaults "yet"), we have

(7.1)
$$\mathbb{P}\left(\tau_1 \geq t_1, \ldots, \tau_k \geq t_k \mid \mathcal{G}_t\right) = E_t\left(e^{-\int_t^{t(k)} \mu(s)\, ds}\right),$$

where

(7.2)
$$\mu(t) = \sum_{\{i:t(i)>t\}} \lambda_i(t).$$

Now, in order to place the joint survivorship calculation into a computationally tractable setting, we suppose that the driving filtration $\{\mathcal{F}_t : t \geq 0\}$ is that generated by an affine process X, and that $\lambda_i(t) = \Lambda_i(X_{t-})$, for an affine function $\Lambda_i(\cdot)$ on the state space D of X. The state X_t could include industry or economy-wide business-cycle variables, or market yield-spread information, as well as firm-specific data. In this case, $\mu(t) = M(X_{t-}, t)$, where,

(7.3)
$$M(x, t) = \sum_{\{i:t(i)>t\}} \Lambda_i(x).$$

Because $\Lambda_i(\cdot)$ is affine for each i, so is $M(\cdot, t)$ for each t. Thus, beginning our calculation at time $t(k-1)$, and working recursively backward using the law of iterated expectations, we have, at each $t(i)$, a solution of the form

(7.4)
$$E_{t(i)}\left(e^{-\int_{t(i)}^{t(i+1)} M(X(s),s))\, ds} e^{\alpha(i+1)+\beta(i+1)\cdot X(t(i+1))}\right) = e^{\alpha(i)+\beta(i)\cdot X(t(i))},$$

where the solution coefficients $\alpha(i)$ and $\beta(i)$ are obtained from the generalized Riccati Equation associated with the discount rate $M(\cdot, s)$, a fixed[7] affine function for $t(i) \leq s \leq t(i+1)$. By taking $t(0) = t$, we thus have

(7.5)
$$\mathbb{P}\left(\tau_1 > t_1, \ldots, \tau_k > t_k \mid \mathcal{G}_t\right) = e^{\alpha(0)+\beta(0)\cdot X(t)}.$$

Related calculations are explored in Duffie [1998b]. Applications include portfolio credit risk calculations and collateralized debt obligations (see Duffie and Gârleanu [2001]), credit-linked notes based on the first to default, loans guaranteed by a defaultable guarantor, and default swaps signed by a defaultable counterparty.

---

[7]One could also solve in one step with a generalized Riccato equation having time-dependent coefficients, but in practice one might prefer to use a time-homogeneous affine model for which the solution coefficients $\alpha(i)$ and $\beta(i)$ are known explicitly, arguing for the recursive calculation of the joint survival probability.

8. – Credit Derivatives

We end the course with some examples of credit-derivative pricing, beginning with the most basic and popular, the default swap. We then turn to credit guarantees, spread options on defaultable bonds, irrevocable lines of credit, and ratings-based step-up bonds. For more examples and analysis of credit derivatives, see Chen and Sopranzetti [1999], Cooper and Martin [1996], Davis and Mavroidis [1997], Duffie [1998b], Duffie and Singleton [2002], Longstaff and Schwartz [1995b], Pierides [1997], and Schönbucher [2003b].

8.1. – Default Swaps

The simplest form of credit derivative is a default swap, Also known (redundantly) as a "credit default swap" (CDS), which pays the buyer of protection, at the default time of a stipulated loan or bond, the difference between its face value and its recovery value, provided the default occurs before a stated expiration date T. The buyer of protection makes periodic coupon payments of some amount U each, until default or T, whichever is first. This is, in effect, an insurance contract for the event of default.

The initial pricing problem is to determine the credit default swap rate (CDS rate) U, normally expressed at an annualized rate. For example, semi-annual payments at a rate of $U = 0.03$ per unit of face value means a CDS rate of 6%. Some discussion and contractual details are provided in Duffie and Singleton [2002].

Given a short-term interest rate process r and fixing an equivalent martingale measure $\mathbb{Q}$, the total market value of the protection offered, per unit of face value, is

$$(8.1) \qquad B = E^{\mathbb{Q}}\left(e^{-\int_0^{\min(T,\tau)} r(u)\,du}(1 - W)1_{\{\tau < T\}}\right),$$

where W is the recovery per unit of face value.

The market value of the coupon payments made by the buyer of protection is

$$(8.2) \qquad A = U \sum_{i=1}^{n} V(t_i)$$

where $t_1, \ldots, t_n = T$ are the coupon dates for the default swap and

$$(8.3) \qquad V(t) = E^{\mathbb{Q}}\left(e^{-\int_0^t r(u)\,du}1_{\{\tau > t\}}\right)$$

is the price of a zero-coupon no-recovery bond whose maturity date is t.

For the default-swap contract to be of zero market value to each counterparty, it must be the case that $A = B$, and therefore that

$$(8.4) \qquad U = \frac{B}{\sum_{i=1}^{n} V(t_i)}.$$

For example, suppose that we are in the setting of Theorem 2, and that the underlying bond has a risk-neutral default intensity process $\lambda^{\mathbb{Q}}$. Then

$$(8.5) \qquad V(t) = E^{\mathbb{Q}}\left(e^{-\int_0^t [r(s)+\lambda^{\mathbb{Q}}(s)]\,ds}\right)$$

and, based on the same calculations used in Section 6,

$$(8.6) \qquad B = \int_0^T q(t)\,dt,$$

where

$$(8.7) \qquad q(t) = E^{\mathbb{Q}}\left[e^{-\int_0^t [\lambda^{\mathbb{Q}}(s)+r(s)]\,ds}\lambda^{\mathbb{Q}}(t)(1 - w(t))\right],$$

and where $w(t)$, is the expected recovery conditional on information available at time t, assuming that default is about to occur, in the sense defined more carefully in Section 6. In practice, $w(t)$ is often taken to be a constant risk-neutral mean recovery level, although attention is increasingly paid to the empirically relevant case of negative correlation between recovery and default intensity.

In an affine setting, $V(t)$, $q(t)$, and thus the credit default swap rate U of (8.4) can be calculated routinely.

8.2. – Credit Guarantees

We will suppose that a loan to a defaultable borrower with a default time τ_B has a guarantor whose default time is τ_G. We make the assumption that, in the event of default by the borrower before the maturity date T of the loan, the guarantor simply takes over the obligation to pay the notional amount of the loan at the original maturity date. This is somewhat unrealistic, but simplifies the exposition. The guaranteed loan thus defaults, in effect, once both the borrower and guarantor default, if both do before the loan's maturity. In practice, the contractual obligation of the guarantor is normally to pay the full principal on the loan within a short time period after the borrower defaults. Our modeled price of the guaranteed loan is therefore conservative, that is, lowered by the assumption that the guarantor may delay paying the obligation until the original maturity date of the loan.

For any given date t, the event that at least one of the borrower and the guarantor survive until t is $A_t = \{\tau_B > t\} \cup \{\tau_G > t\}$ that at least one survives to t. We have

$$(8.8) \qquad \mathbb{Q}(A_t) = \mathbb{Q}(\{\tau_B > t\}) + \mathbb{Q}(\{\tau_G > t\}) - \mathbb{Q}(\{\tau_B > t\} \cap \{\tau_G > t\}).$$

We will assume, in order to obtain concrete calculations, a doubly stochastic model for τ_B and τ_G, with respective risk-neutral intensities λ^B and λ^G. From (8.8),

$$(8.9) \qquad \begin{aligned} \mathbb{Q}(A_t) &= \mathbb{E}^{\mathbb{Q}}\left(e^{-\int_0^t \lambda^B(s)\,ds}\right) + \mathbb{E}^{\mathbb{Q}}\left(e^{-\int_0^t \lambda^G(s)\,ds}\right) \\ &\quad - \mathbb{E}^{\mathbb{Q}}\left(e^{-\int_0^t [\lambda^B(s)+\lambda^G(s)]\,ds}\right), \end{aligned}$$

each term of which can be easily calculated if both intensities, λ^B and λ^G, are affine with respect to a state process X that is affine under $\mathbb{Q}$.

From this, depending on the recovery model and the probabilistic relationship between interest rates and default times, one has relatively straightforward pricing of the guaranteed loan. For example, suppose that interest rates are independent under $\mathbb{Q}$ of the default times τ_B and τ_G, and assume constant recovery of a fraction w of the face value of the loan. We let $\delta(t) = E^{\mathbb{Q}}(e^{-\int_0^t r(u)\,du})$ denote the price of a default-free zero-coupon bond of maturity t, of unit face value. Then the market value of the guaranteed loan, per unit of face value, is

$$(8.10) \qquad V = \delta(T)q(T) - w \int_0^T \delta(t)q'(t)\,dt,$$

where $q(t) = \mathbb{Q}(A_t)$ is the risk-neutral probability that at least one of the borrower and the guarantor survive to t, so that $-q'(t)$ is the risk-neutral density of the time of default of the guaranteed loan. In an affine setting, $q'(t)$ is easily computed, so the computation of (8.10) is straightforward.

Many variations are possible, including a default recovery from the guarantor differing from that of the original borrower.

8.3. – Spread Options

The yield of a zero-coupon bond of unit face value, of price V, and of maturity t is $(-\log V)/t$. The yield spread of one bond relative to another of a lower yield is simply the difference in their yields. The prices of defaultable bonds are often quoted in terms of their yield spreads relative to some benchmark, such as government bonds. (Swap rates are often used as a benchmark, but we defer that issue to Duffie and Singleton [2002].)

We will consider the price of an option to put (that is, to sell) at a given time t a defaultable zero-coupon bond at a given spread $\bar{s}$ relative to a benchmark yield $Y(t)$. The remaining maturity of the bond at time t is m. This option therefore has a payoff at time t of

$$(8.11) \qquad Z = \left(e^{-(Y(t)+\bar{s})m} - e^{-(Y(t)+S(t))m} \right)^+,$$

where $S(t)$ is the spread of the defaultable bond at time t. For simplicity, we will suppose that a contractual knockout provision prevents exercise of the option in the event that default occurs before the exercise date t. A slightly more complicated result arises if the option can be exercised even after the default of the underlying bond. We will also suppose for convenience that the benchmark yield $Y(t)$ is the default-free yield of the same time m to maturity. The market value of this spread option is therefore

$$(8.12) \qquad V = E^{\mathbb{Q}}\left[e^{-\int_0^t r(u)\,du} 1_{\{\tau > t\}} Z \right].$$

We will consider a setting with a constant fractional loss ℓ of market value at default, as in Section 6.3, so that the defaultable bond has a price at time t, on the event that $\tau > t$, of

$$(8.13) \qquad e^{-(Y(t)+S(t))m} = E_t^{\mathbb{Q}}\big(e^{-\int_t^{t+m} R(u)\,du}\big),$$

where $R(u) = r(u) + \ell\lambda^{\mathbb{Q}}(u)$ is the default-adjusted short rate.

In order to simplify the calculations, we will suppose that τ is doubly stochastic driven by a state process X that is affine under $\mathbb{Q}$, with a risk-neutral intensity $\lambda_t^{\mathbb{Q}} = a_0 + a_1 \cdot X_{t-}$. We also suppose that the default-free short-rate process r is affine with respect to X, so that $r_t = \gamma_0 + \gamma_1 \cdot X_t$, and the default-free reference yield to maturity m is of thus of the form $Y(t) = \theta_0 + \theta_1 \cdot X(t)$.

The default adjusted short rate is $R_t = \rho_0 + \rho_1 \cdot X_t$, for $\rho_0 = \gamma_0 + \ell a_0$ and $\rho_1 = \gamma_1 + \ell a_1$, so we have a defaultable yield spread, in the event of survival to t, of the form $S(t) = \xi_0 + \xi_1 \cdot X(t)$. The option payoff, in the event of no default by t, is thus of the form

$$(8.14) \qquad Z = \big(e^{c_0 + c_1 \cdot X(t)} - e^{f_0 + f_1 \cdot X(t)}\big) 1_{\{d \cdot X \le y\}},$$

where

$$f_0 = (\theta_0 + \xi_0)m$$
$$f_1 = (\theta_1 + \xi_1)m$$
$$c_0 = (\theta_0 + \bar{s})m$$
$$c_1 = \theta_1 m$$
$$d = f_1 - c_1$$
$$y = c_0 - f_0.$$

From (8.12) and the doubly-stochastic assumption, we therefore have $V(t) = V_0(t) + V_1(t)$, where

$$(8.15) \qquad V_0(t) = E^{\mathbb{Q}}\big[e^{-\int_0^t [\zeta_0 + \zeta_1 \cdot X(t)]\,du} e^{c_0 + c_1 \cdot X(t)} 1_{\{d \cdot X \le y\}}\big]$$

$$(8.16) \qquad = G_{t,\zeta,c,d}(y)$$

$$(8.17) \qquad V_1(t) = E^{\mathbb{Q}}\big[e^{-\int_0^t [\zeta_0 + \zeta_1 \cdot X(t)]\,du} e^{f_0 + f_1 \cdot X(t)} 1_{\{d \cdot X \le y\}}\big],$$

$$(8.18) \qquad = G_{t,\zeta,f,d}(y)$$

where $\zeta_0 = \gamma_0 + a_0$, $\zeta_1 = \gamma_1 + a_1$, and

$$(8.19) \qquad G_{t,\zeta,c,d}(y) = E^{\mathbb{Q}}\big[e^{-\int_0^t \zeta_0 + \zeta_1 \cdot X(t)} e^{c_0 + c_1 \cdot X(t)} 1_{\{d \cdot X(t) \le y\}}\big].$$

Using the approach of Stein and Stein [1991] and Heston [1993] for option pricing, $G_{t,\zeta,c,d}(\cdot)$ can be calculated by inverting its Fourier Transform $\hat{G}_{t,\zeta,c,d}(\cdot)$, which is defined by

$$\hat{G}_{t,\zeta,c,d}(z) = \int_{\mathbb{R}} e^{izy}\,dG_{t,\zeta,c,d}(y)$$

$$= E^{\mathbb{Q}}\big[e^{-\int_0^t \zeta_0 + \zeta_1 \cdot X(t)} e^{c_0 + (c_1 + izd) \cdot X(t)}\big],$$

using Fubini's theorem.

That is, under regularity, we can apply the Lévy inversion formula,

$$G_{t,\zeta,c,d}(y) = \frac{\hat{G}_{t,\zeta,c,d}(0)}{2} - \frac{1}{\pi} \int_0^\infty \frac{\text{Im}\left[\hat{G}_{t,\zeta,c,d}(z)e^{-izy}\right]}{z} \, dz,$$

where $\text{Im}(w)$ denotes the imaginary part of any complex number w.

The affine model is convenient, for it provides an easily computed solution of the Fourier transform of the form

(8.20) $$\hat{G}_{t,\zeta,c,d}(z) = e^{\hat{\alpha}(t;z)+\hat{\beta}(t;z)\cdot X(0)},$$

where we have indicated explicitly the dependence on z of the boundary condition of the generalized Riccati equations for $\hat{\alpha}(t; z)$ and $\hat{\beta}(t, z)$.

In practice, one might chooses a parameterization of X for which $\hat{\alpha}$ and $\hat{\beta}$ are explicit (for example multivariate versions of the basic affine model), or one may solve the generalized Riccati equations by a numerical method, such as Runge-Kutta.

8.4. – Irrevocable Lines of Credit

Banks often provide a credit facility under which a borrower is offered the option to enter into short-term loans for up to a given notional amount N, until a given time T, all at a contractually fixed spread. The short-term loans are of some term m. That is, at any time t among the potential borrowing dates $m, 2m, 3m, \ldots, T$, the borrower may enter into a loan maturing at time $t + m$, of maximum size N, at a fixed spread $\bar{s}$ over the current reference yield $Y(t)$ for m-period loans.

Such a credit facility may be viewed as a portfolio of defaultable put options sold to the borrower on the borrower's own debt. The borrower is usually charged a fee, however, for the unused portion of the credit facility, say F per unit of unused notional.

Our objective is to calculate the market value to the borrower of the credit facility, including the effect of any fees paid. We will ignore the fact that use of the credit facility may either signal or affect the borrower's credit quality. This endogeneity is not easily treated directly by the option pricing method that we will use. Instead, we will assume the same doubly-stochstic affine model of default given in the previous application to defaultable bond options. We will also assume that the bank is default free. (Otherwise, access to the facility would be somewhat less valuable.) We also ignore legal and other institutional impediments to the use of facility that can be important in practice, and that often lead to only partial use of the facility, despite the fact that in our model setting the facility is either fully utilized by the borrower at a point in time, or not used at all.

For each unit of notional, the option to use the facility at time t actually means the option to sell, at a price equal to 1, a bond promising to pay the bank $e^{(Y(t)+\bar{s})m}$ at time $t + m$. The market value of this obligation at time t, assuming that the borrower has survived to time t, is therefore

$$e^{-(Y(t)+S(t))m}e^{(Y(t)+\bar{s})m} = e^{(\bar{s}-S(t))m},$$

recalling that $S(t)$ is the borrower's spread on loans at time t maturing at time $t + m$.

If surviving to time t, the value of the option to the borrower, per unit of notional, reflecting also the benefit of the reduction in the fee F per unit of unused credit, is, in the affine setting of the previous treatment of defaultable debt options,

(8.21)
$$\begin{aligned}
H(t) &= \left(1 + F - e^{(\bar{s}-S(t))m}\right)^+ \\
&= \left(1 + F - e^{h_0+h_1 \cdot X(t)}\right)1_{g \cdot X(t) \le v},
\end{aligned}$$

where $h_0 = (\bar{s} - \xi_0)m$, $h_1 = -m\xi_1$, $g = -m\xi_1$, and $v = \log(1 + F) + \xi_0 - \bar{s}$. Here, we used our previous spread calculation, $S(t) = \xi_0 + \xi_1 \cdot X(t)$.

The optionality value of the credit facility for the portion of the period beginning at time t is

$$\begin{aligned}
U(t) &= E^{\mathbb{Q}}\left[e^{-\int_0^t r(s)\,ds}H(t)1_{\{\tau > t\}}\right] \\
&= E^{\mathbb{Q}}\left[e^{-\int_0^t [r(s)+\lambda^{\mathbb{Q}}(s)]\,ds}H(t)\right] \\
&= E^{\mathbb{Q}}\left[e^{-\int_0^t [\zeta_0+\zeta_1 \cdot X(s)]\,ds}\left(1 + F - e^{h_0+h_1 \cdot X(t)}\right)1_{\{g \cdot X(t) \le v\}}\right] \\
&= (1 + F)G_{t,\zeta,0,g}(v) - G_{t,\zeta,h,g}(v).
\end{aligned}$$

We can therefore calculate $U(t)$ by the Fourier inversion method of the previous application to defaultable debt options.

The fee F is paid at time t only when the borrower has not defaulted by t, and only when the borrower is not drawing on the facility at t. We have already accounted in our calculation of $U(t)$ for the benefit of not paying the fee when the facility is used, so the total market value of the facilty to the borrower is

(8.22)
$$V(F) = \sum_{i=1}^{T/m} U(mi) - F e^{\alpha(mi)+\beta(mi) \cdot X(0)},$$

where $\alpha(\cdot)$ and $\beta(\cdot)$ solve the the generalized Riccati equation associated with the survival-contingent discount

(8.23)
$$E^{\mathbb{Q}}\left(e^{-\int_0^t [\zeta_0+\zeta_1 \cdot X(s)]\,ds}\right) = e^{\alpha(t)+\beta(t) \cdot X(0)}.$$

If the credit facility is priced on a stand-alone zero-profit basis, the associated fee F would be set to solve the equation $V(F) = 0$. In practice, the fee is not necessarily set in this fashion. On May 3, 2002, for example, *The Financial Times* indicated concern by banks over their pricing policy on credit facilities, and indicated fees on undrawn lines had recently been ranging from roughly 9 basis points for A-rated firms to roughly 20 basis points for BBB-rated firms. Banks are also, apparently, beginning to build some protection against the optionality in the exercise of these lines, by using fees (F) or spreads $(\bar{s})$ that depend on the fraction of the line that is drawn.

8.5. – Ratings-Based Step-Up Bonds

Most publicly traded debt, and much privately issued debt, is assigned a credit rating, essentially a credit quality score, by one or more of the major credit rating agencies. For example, the standard letter credit ratings for Moodys are Aaa, Aa, A, Baa, Ba, B, and D (for default). There are 3 refined ratings for each letter rating, as in Ba1, Ba2, and Ba3. The term "investment grade" means "rated Aaa, Aa, A, or Baa." For Moodys, speculative-grade ratings are those below Baa. For Standard and Poors, whose letter ratings are AAA, AA, A, BBB, BB, B, C, and D, "speculative-grade" means below BBB.

It has become increasingly common for bond issuers to link the size of the coupon rate on their debt with their credit rating, offering a higher coupon rate at lower ratings, perhaps in an attempt to appeal to investors based on some degree of hedging against a decline in credit quality. This embedded derivative is called a "ratings-based step-up." For example, *The Financial Times* reported on April 9, 2002, that a notional amount of approximately 120 billion Euros of such ratings-based step-up bonds had been issued by telecommunications firms, one of which, a 25-billion-Euro Deutsche Telekom bond, would begin paying at extra 50 basis points (0.5%) in interest per year with the downgrade by Standard and Poors of Deutsche Telekom debt from A− to BBB+ on April 8, 2002, following[8] a similar downgrade by Moodys.

There is a potentially adverse effect of such a step-up feature, however, for a downgrade brings with it an additional interest expense, which, depending on the capital structure and cash flow of the issuer, may actually reduce the total market value of the debt, and even bring on further ratings downgrades, higher coupon rates, and so on. Manso, Strulovici, and Tchistyi [2003] characterize this effect, and demonstrate the inefficiency of step-up debt relative to straight debt. We ignore this feedback effect in our calculation of the pricing of ratings-based step-up bonds.

We simplify the pricing problem by considering the common case in which the only ratings changes that cause a change in coupon rate are into and out of the investment-grade ratings categories. That is, we assume that the coupon rate is c_A whenever the issuer's rating is investment grade, and that the speculative-grade coupon rate is $c_B > c_A$.

---

[8]Most ratings-based step-ups occur with a stipulated reduction in rating by any of the major ratings agencies, but this particular bond required a downgrade by both Moodys and Standard and Poors before the step-up provision could occur.

Our pricing model is the same doubly-stochastic model used in earlier applications. In order to treat ratings transitions risk, we will assume that that the risk-neutral default intensity $\lambda_t^{\mathbb{Q}}$ is higher for speculative grade ratings than for investment grade ratings. This is natural. In practice, however, the maximum yield spread associated with investment grade fluctuates with uncertainty over time. There is moreover a momemtum effect in ratings, measured by Lando and Skødeberg [2002], that we do not capture by mapping ratings to intervals of risk-neutral default intensity. In order to maintain tractability, we will suppose that the issuer has an investment-grade rating whenever $\lambda_t^{\mathbb{Q}} \geq \Theta_t$, and is otherwise of speculative grade, where $\Theta(t) = \theta_0 + \theta_1 \cdot X(t-)$. It would be equivalent for purposes of tractability, since yield spreads are affine in $X(t)$ in this setting, to suppose that the maximal level of investment-grade straight-debt yield spreads at a given maturity is affine with respect to $X(t)$.

We will use, for simplicity, a model with zero recovery of coupons at default, and recovery of a given risk-neutral expected fraction w of principal at default.

The coupon paid at coupon date t, in the event of survival to that date, is

(8.24)
$$\begin{aligned}
c(t) &= c_A + (c_B - c_A)1_{\{\lambda^{\mathbb{Q}}(t) \geq \Theta(t)\}} \\
&= c_A + (c_B - c_A)1_{\{h \cdot X(t-) \leq u\}},
\end{aligned}$$

where $h = \theta_1 - a_1$ and $u = a_0 - \theta_0$. The initial market value of this coupon is

$$\begin{aligned}
F(t) &= E^{\mathbb{Q}}\left[e^{-\int_0^t r(s)\,ds} c(t)1_{\{\tau > t\}}\right] \\
&= E^{\mathbb{Q}}\left[e^{-\int_0^t [r(s) + \lambda^{\mathbb{Q}}(s)]\,ds} c(t)\right] \\
&= E^{\mathbb{Q}}\left[e^{-\int_0^t [\zeta_0 + \zeta_1 \cdot X(s)]\,ds}[c_A + (c_B - c_A)1_{\{h \cdot X(t-) \leq u\}}]\right], \\
&= c_A e^{\alpha(t) + \beta(t) \cdot X(0)} + (c_B - c_A)G_{t,\zeta,0,h}(u),
\end{aligned}$$

where $e^{\alpha(t) + \beta(t) \cdot X(0)} = E^{\mathbb{Q}}\left(e^{-\int_0^t [\zeta_0 + \zeta_1 \cdot X(s)]\,ds}\right)$. Calculation of $G_{t,\zeta,0,h}(u)$ is by the Fourier inversion method used previously.

For principal payment date T and coupon dates $t_1, t_2, \ldots, t_n = T$, the initial price of the ratings-based step-up bond, under our assumptions, is then

(8.25)
$$V_0 = e^{\alpha(T) + \beta(T) \cdot X(0)} + \int_0^T \Phi(0, t)\,dt + \sum_{i=1}^n F(t_i),$$

where $\Phi(0, t)$ is the market-value density for the recovery of principal, calculated in an affine setting as in (6.8).

Appendices

A. – Structural Models of Default

This appendix, which draws from Chapter 11 of Duffie [2001], reviews the most basic classes of structural models of default risk, which are built on a direct model of survival based on the sufficiency of assets to meet liabilities.

For this appendix, we let B be a standard Brownian motion in $\mathbb{R}^d$ on a complete probability space $(\Omega, \mathcal{F}, \mathbb{P})$, and we fix the standard filtration $\{\mathcal{F}_t : t \geq 0\}$ of B.

A.1. – The Black-Scholes-Merton Model

For the Black-Scholes-Merton model, based on Black and Scholes [1973] and Merton [1974], we may think of equity and debt as derivatives with respect to the total market value of the firm, and priced accordingly. In the literature, considerable attention has been paid to market imperfections and to control that may be exercised by holders of equity and debt, as well as managers. With these market imperfections, the theory becomes more complex and less like a derivative valuation model.

With the classic Black-Scholes-Merton model of corporate debt and equity valuation, one supposes that the firm's future cash flows have a total market value at time t given by A_t, where A is a geometric Brownian motion, satisfying

$$dA_t = \varphi A_t \, dt + \sigma A_t \, dB_t,$$

for constants φ and $\sigma > 0$, and where we have taken $d = 1$ as the dimension of the underlying Brownian motion B. One sometimes refers to A_t as the assets of the firm. We will suppose for simplicity that the firm produces no cash flows before a given time T. In order to justify this valuation of the firm, one could assume there are other securities available for trade that create the effect of complete markets, namely that, within the technical limitations of the theory, any future cash flows can be generated as the dividends of a trading strategy with respect to the available securities. There is then a unique price at which those cash flows would trade without allowing an arbitrage.

We take it that the original owners of the firm have chosen a capital structure consisting of pure equity and of debt in the form of a single zero-coupon bond maturing at time T, of face value L. In the event that the total value A_T of the firm at maturity is less than the contractual payment L due on the debt, the firm defaults, giving its future cash flows, worth A_T, to debtholders. That is, debtholders receive $\min(L, A_T)$ at T. Equityholders receive the residual $\max(A_T - L, 0)$. We suppose for simplicity that there are no other distributions (such as dividends) to debt or equity. We will shortly confirm the natural conjecture that the market value of equity is given by the Black-Scholes option-pricing formula, treating the firm's asset value as the price of the underlying security.

Bond and equity investors have already paid the original owners of the firm for their respective securities. The absence of well-behaved arbitrage implies that at, any time $t < T$, the total of the market values S_t of equity and Y_t of debt must be the market value A_t of the assets. This is one of the main points made by Modigliani and Miller [1958], in their demonstration of the irrelevance of capital structure in perfect markets.

Markets are complete given riskless borrowing or lending at a constant rate r and given access to a self-financing trading strategy whose value process is A. (See, for example, Duffie [2001], Chapter 6.) This implies that there is at most one equivalent martingale measure.

Letting $B_t^Q = B_t + \eta t$, where $\eta = (\varphi - r)/\sigma$, we have

$$dA_t = rA_t\,dt + \sigma A_t\,dB_t^Q.$$

Girsanov's theorem states that B^Q is a standard Brownian motion under the equivalent probability measure $\mathbb{Q}$ defined by

$$\frac{d\mathbb{Q}}{d\mathbb{P}} = e^{-\eta B(T) - \eta^2 T/2}.$$

By Ito's formula, $\{e^{-rt}A_t : t \in [0, T]\}$ is a $\mathbb{Q}$-martingale. It follows that, after discounting by e^{-rt}, $\mathbb{Q}$ is the equivalent martingale measure. As $\mathbb{Q}$ is unique in this regard, we have the unique price process S of equity in the absence of well-behaved arbitrage (see, for example, Duffie [2001], Chapter 6), given by

$$S_t = E_t^{\mathbb{Q}}\left[e^{-r(T-t)}\max(A_T - L, 0)\right].$$

Thus, the equity price S_t is computed by the Black-Scholes option-pricing formula, treating A_t as the underlying asset price, σ as the volatility coefficient, the face value L of debt as the strike price, and $T - t$ as the time remaining to exercise. The market value of debt at time t is the residual, $A_t - S_t$.

When the original owners of the firm sold the debt with face value L and the equity, they realized a total initial market value of $S_0 + Y_0 = A_0$, which does not depend on the chosen face value L of debt. This is again an aspect of the Modigliani-Miller theorem. The same irrelevance of capital structure for the total valuation of the firm applies much more generally, and has nothing to do with geometric Brownian motion, nor with the specific nature of debt and equity. With market imperfections, however, the design of the capital structure can be important in this regard.

Fixing the current value A_t of the assets, the market value S_t of equity is increasing in the asset volatility parameter σ, due to the usual Jensen effect in the Black-Scholes formula. Thus, equity owners, were they to be given the opportunity to make a switch to a "riskier technology," one with a larger asset volatility parameter, would increase their market valuation by doing so, at the expense of bondholders, provided the total initial market value of the firm is not reduced too much by the switch. This is a simple example of what is sometimes called "asset substitution."

Given the time value of the option embedded in equity, bondholders would prefer to advance the maturity date of the debt; equityholders would prefer to extend it.

Equityholders (or managers acting as their agents) typically hold the power to make decisions on behalf of the firm, subject to legal and contractual restrictions such as debt covenants. This is natural in light of equity's position as the residual claim on the firm's cash flows.

Geske [1977] used compound option modeling so as to extend to debt at various maturities.

A.2. – First-Passage Models of Default

A "first-passage" model of default is one for which the default time is the first time that the market value of the assets of the issuer have reached a sufficiently low level. Black and Cox [1976] developed the idea of first-passage-based default timing, but used an exogenous default boundary. We shift now to a slightly more elaborate setting for the valuation of debt and equity, and consider the endogenous timing of default, using an approach formulated by Fisher, Heinkel, and Zechner [1989] and solved and extended by Leland [1994], and subsequently, by others.

We take as given an equivalent martingale measure $\mathbb{Q}$. (In this infinite-horizon setting, by an equivalent martingale measure, we require only that, for each finite t, $\mathbb{Q}$ and $\mathbb{P}$ equivalent when restricted to $\mathcal{F}_t$.)

The resources of a given firm are assumed to consist of cash flows at the rate δ_t for each time t. We suppose that δ is an adapted process with $\int_0^t |\delta_s|\, ds < \infty$ almost surely for all t. The market value of the assets of the firm at time t is defined as the market value A_t of the future cash flows.

That is,

(A.1)
$$A_t = E_t^{\mathbb{Q}} \left[\int_t^\infty e^{-r(s-t)} \delta_s \, ds \right].$$

We assume that A_t is well defined and finite for all t. The martingale representation theorem, which also applies under $\mathbb{Q}$ for the Brownian motion $B^{\mathbb{Q}}$, then implies that

(A.2)
$$dA_t = (rA_t - \delta_t) \, dt + \sigma_t \, dB_t^{\mathbb{Q}},$$

where σ is an adapted $\mathbb{R}^d$-valued process such that $\int_0^T \sigma_t \cdot \sigma_t \, dt < \infty$ for all $T \in [0, \infty)$, and where $B^{\mathbb{Q}}$ is the standard Brownian motion in $\mathbb{R}^d$ under $\mathbb{Q}$ obtained from B and Girsanov's theorem.

We suppose that the original owners of the firm chose its capital structure to consist of a single bond as its debt, and pure equity, defined in detail below. The bond and equity investors have already paid the original owners for these securities. Before we consider the effects of market imperfections, the total of the market values of equity and debt must be the market value A of the assets, which is a given process, so the design of the capital structure is again irrelevant from the viewpoint of maximizing the total value received by the original owners of the firm.

For simplicity, we suppose that the bond promises to pay coupons at a constant total rate c, continually in time, until default. This sort of bond is sometimes called a consol. Equityholders receive the residual cash flow in the form of dividends at the rate $\delta_t - c$ at time t, until default. At default, the firm's future cash flows are assigned to debtholders.

The equityholders' dividend rate, $\delta_t - c$, may have negative outcomes. It is commonly stipulated, however, that equity claimants have limited liability, meaning that they should not experience negative cash flows. One can arrange for limited liability by dilution of equity. That is, so long as the market value of equity remains strictly positive, newly issued equity can be sold into the market so as to continually finance the negative portion $(c - \delta_t)^+$ of the residual cash flow. (Alternatively, the firm could issue debt, or other forms of securities, to finance itself.) When the price of equity reaches zero, and the financing of the firm through equity dilution is no longer possible, the firm is in any case in default, as we shall see. While dilution increases the quantity of shares outstanding, it does not alter the total market value of all shares, and so is a relatively simple modeling device. Moreover, dilution is irrelevant to individual shareholders, who would in any case be in a position to avoid negative cash flows by selling their own shares as necessary to finance the negative portion of their dividends, with the same effect as if the firm had diluted their shares for this purpose. We are ignoring here any frictional costs of equity issuance or trading. This is another aspect of the Modigliani-Miller theory, the irrelevance of dividend policy.

Equityholders are assumed to have the contractual right to declare default at any stopping time T, at which time equityholders give up to debtholders the rights to all future cash flows, a contractual arrangement termed strict priority, or sometimes absolute priority. We assume that equityholders are not permitted to delay liquidation after the value A of the firm reaches 0, so we ignore the possibility that $A_T < 0$. We could also consider the option of equityholders to change the firm's production technology, or to call in the debt for some price.

The bond contract conveys to debtholders, under a protective covenant, the right to force liquidation at any stopping time τ at which the asset value A_τ is as low or lower than some stipulated level, which we take for now to be the face value L of the debt. Debtholders would receive A_τ at such a time τ; equityholders would receive nothing.

Assuming that $A_0 > L$, we first consider the total coupon payment rate c that would be chosen at time 0 in order that the initial market value of the bond is its face value L. Such a bond is said to be "at par," and the corresponding coupon rate per unit of face value, c/L, is the par yield. If bondholders rationally enforce their protective covenant, we claim that the par yield must be the riskless rate r. We also claim that, until default, the bond paying coupons at the total rate $c = rL$ is always priced at its face value L, and that equity is always priced at the residual value, $A - L$. Finally, equityholders have no strict preference to declare default on a par-coupon bond before $\tau(L) = \inf\{t : A_t \leq L\}$, which is the first time allowed for in the protective covenant, and bondholders rationally force liquidation at $\tau(L)$.

If the total coupon rate c is strictly less than the par rate rL, then equityholders never gain by exercising the right to declare default (or, if they have it, the right to call the debt at its face value) at any stopping time T with $A_T \geq L$, because the market value at time T of the future cash flows to the bond is strictly less than L if liquidation occurs at a stopping time $U > T$ with $A_U \leq L$. Avoiding liquidation at T would therefore leave a market value for equity that is strictly greater than $A_T - L$. With $c < rL$, bondholders would liquidate at the first time $\tau(L)$ allowed for in their protective covenant, for by doing so they receive L at $\tau(L)$ for a bond that, if left alive, would be worth less than L. In summary, with $c < rL$, the bond is liquidated at $\tau(L)$, and trades at a "discount" price at any time t before liquidation, given by

$$(\text{A.3}) \qquad Y_t = E_t^Q\left[\int_t^{\tau(L)} e^{-r(s-t)}c\,ds + e^{-r(\tau(L)-t)}L\right]$$

$$(\text{A.4}) \qquad = \frac{c}{r} + E_t^Q[e^{-r(\tau(L)-t)}]\left(L - \frac{c}{r}\right) < L.$$

A.3. – Example: Brownian Dividend Growth

As an example, suppose the cash-flow rate process δ is a geometric Brownian motion under Q, in that

$$d\delta_t = \mu\delta_t\,dt + \sigma\delta_t\,dB_t^Q,$$

for constants μ and σ, where B^Q is a standard Brownian motion under Q. We assume throughout that $\mu < r$, so that, from (A.1), A is finite and

$$dA_t = \mu A_t \, dt + \sigma A_t \, dB_t^Q.$$

We calculate that $\delta_t = (r - \mu)A_t$.

For any given constant $K \in (0, A_0)$, the market value of a security that claims one unit of account at the hitting time $\tau(K) = \inf\{t : A_t \leq K\}$ is, at any time $t < \tau(K)$,

(A.5)
$$E_t^Q\left[e^{-r(\tau(K)-t)}\right] = \left(\frac{A_t}{K}\right)^{-\gamma},$$

where

(A.6)
$$\gamma = \frac{m + \sqrt{m^2 + 2r\sigma^2}}{\sigma^2},$$

and where $m = \mu - \sigma^2/2$. One can verify (A.5) as an exercise, applying Itô's formula.

Let us consider for simplicity the case in which bondholders have no protective covenant. Then equityholders declare default at a stopping time that solves the maximum equity valuation problem

(A.7)
$$w(A_0) \equiv \sup_{T \in \mathcal{T}} \; E^Q\left[\int_0^T e^{-rt}(\delta_t - c)\, dt\right],$$

where $\mathcal{T}$ is the set of stopping times.

We naturally conjecture that the maximization problem (A.7) is solved by a hitting time of the form $\tau(A_B) = \inf\{t : A_t \leq A_B\}$, for some default-triggering level A_B of assets, to be determined. Given this conjecture, we further conjecture from Ito's formula that the function $w : (0, \infty) \to [0, \infty)$ defined by (A.7) solves the ODE

(A.8)
$$\mathcal{A}w(x) - rw(x) + (r - \mu)x - c = 0, \quad x > A_B,$$

where

(A.9)
$$\mathcal{A}w(x) = w'(x)\mu x + \frac{1}{2}w''(x)\sigma^2 x^2,$$

with the absolute-priority boundary condition

(A.10)
$$w(x) = 0, \quad x \leq A_B.$$

Finally, we conjecture the smooth-pasting condition

(A.11)
$$w'(A_B) = 0,$$

based on (A.10) and continuity of the first derivative $w'(\cdot)$ at A_B. Although not an obvious requirement for optimality, the smooth-pasting condition, sometimes called the high-order-contact condition, has proven to be a fruitful method by which to conjecture solutions, as follows.

If we are correct in conjecturing that the optimal default time is of the form $\tau(A_B) = \inf\{t : A_t \leq A_B\}$, then, given an initial asset level $A_0 = x > A_B$, the value of equity must be

$$\text{(A.12)} \qquad w(x) = x - A_B \left(\frac{x}{A_B}\right)^{-\gamma} - \frac{c}{r}\left[1 - \left(\frac{x}{A_B}\right)^{-\gamma}\right].$$

This conjectured value of equity is merely the market value x of the total future cash flows of the firm, less a deduction equal to the market value of the debtholders' claim to A_B at the default time $\tau(A_B)$ using (A.5), less another deduction equal to the market value of coupon payments to bondholders before default. The market value of those coupon payments is easily computed as the present value c/r of coupons paid at the rate c from time 0 to time $+\infty$, less the present value of coupons paid at the rate c from the default time $\tau(A_B)$ until $+\infty$, again using (A.5). In order to complete our conjecture, we apply the smooth-pasting condition $w'(A_B) = 0$ to this functional form (A.12), and by calculation obtain the conjectured default triggering asset level as

$$\text{(A.13)} \qquad\qquad A_B = \beta c,$$

where

$$\text{(A.14)} \qquad\qquad \beta = \frac{\gamma}{r(1 + \gamma)}.$$

We are ready to state and verify Leland's pricing result.

PROPOSITION 1. *The default-timing problem (A.7) is solved by* $\inf\{t : A_t \leq \beta c\}$. *The associated initial market value* $w(A_0)$ *of equity is* $W(A_0, c)$, *where*

$$\text{(A.15)} \qquad\qquad W(x, c) = 0, \quad x \leq \beta c,$$

and

$$\text{(A.16)} \qquad W(x, c) = x - \beta c \left(\frac{x}{\beta c}\right)^{-\gamma} - \frac{c}{r}\left[1 - \left(\frac{x}{\beta c}\right)^{-\gamma}\right], \quad x \geq \beta c.$$

The initial value of debt is $A_0 - W(A_0, c)$.

The following proof, a verification of Leland's solution, is adapted from Duffie and Lando [2001].

PROOF. First, it may be checked by calculation that $W(\cdot, c)$ satisfies the differential equation (A.8) and the smooth-pasting condition (A.11). Ito's formula applies to C^2 (twice continuously differentiable) functions. In our case, although $W(\cdot, c)$ need not be C^2, it is convex, is C^1, and is C^2 except at βc, where $W_x(\beta c, c) = 0$. Under these conditions, we obtain the result, as though from a standard application of Ito's formula,[1]

$$(A.17) \quad W(A_s, c) = W(A_0, c) + \int_0^s AW(A_t, c)\, dt + \int_0^s W_x(A_t, c)\sigma A_t\, dB_t^{\mathbb{Q}},$$

where

$$(A.18) \qquad AW(x, c) = W_x(x, c)\mu x + \frac{1}{2}W_{xx}(x, c)\sigma^2 x^2,$$

except at $x = \beta c$, where we may replace "$W_{xx}(\beta c, c)$" with zero.

For each time t, let

$$q_t = e^{-rt}W(A_t, c) + \int_0^t e^{-rs}((r - \mu)A_s - c)\, ds.$$

From Ito's formula,

$$(A.19) \qquad dq_t = e^{-rt}f(A_t)\, dt + e^{-rt}W_x(A_t, c)\sigma A_t\, dB_t^{\mathbb{Q}},$$

where

$$f(x) = AW(x, c) - rW(x, c) + (r - \mu)x - c.$$

Because W_x is bounded, the last term of (A.19) defines a $\mathbb{Q}$-martingale. For $x \leq \beta c$, we have both $W(x, c) = 0$ and $(r - \mu)x - c \leq 0$, so $f(x) \leq 0$. For $x > \beta c$, we have (A.8), and therefore $f(x) = 0$. The drift of q is therefore never positive, and for any stopping time T we have $q_0 \geq E^{\mathbb{Q}}(q_T)$, or equivalently,

$$(A.20) \qquad W(A_0, c) \geq E^{\mathbb{Q}}\left[\int_0^T e^{-rs}(\delta_s - c)\, ds + e^{-rT}W(A_T, c)\right].$$

For the particular stopping time $\tau(\beta c)$, we have

$$(A.21) \qquad W(A_0, c) = E^{\mathbb{Q}}\left[\int_0^{\tau(\beta c)} e^{-rs}(\delta_s - c)\, ds\right],$$

---

[1] We use a version of Ito's formula that can be applied to a real-valued function that is C^1 and is C^2 except at a point, as, for example, in [1988], page 219.

using the boundary condition (A.15) and the fact that $f(x) = 0$ for $x > \beta c$. So, for any stopping time T,

(A.22)
$$\begin{aligned} W(A_0, c) &= E^{\mathbb{Q}} \left[\int_0^{\tau(\beta c)} e^{-rs} (\delta_s - c) \, ds \right] \\ &\geq E^{\mathbb{Q}} \left[\int_0^T e^{-rs} (\delta_s - c) \, ds + e^{-rT} W(A_T, c) \right] \\ &\geq E^{\mathbb{Q}} \left[\int_0^T e^{-rs} (\delta_s - c) \, ds \right], \end{aligned}$$

using the nonnegativity of W for the last inequality. This implies the optimality of the stopping time $\tau(\beta c)$ and verification of the proposed solution $W(A_0, c)$ of (A.7). □

This model was further elaborated to treat taxes in Leland [1994], coupon debt of finite maturity in Leland and Toft [1996], endogenous calling of debt and recapitalization in Leland [1998] and Uhrig-Homburg [1998], and incomplete observation of the firmís capital structure by bond investors, with default intensity, in Duffie and Lando [2001]. Yu [2002a] provides empirical support for incomplete observation of the capital structure. Alternative approaches to default recovery are considered by Anderson and Sundaresan [1996], Anderson, Pan, and Sundaresan [1995], Fan and Sundaresan [2000], Mella-Barral [1999], and Mella-Barral and Perraudin [1997].

Longstaff and Schwartz [1995a] developed a similar first-passage defaultable bond pricing model with stochastic default-free interest rates. (See also Nielsen, Saá-Requejo, and Santa-Clara [1993] and Collin-Dufresne and Goldstein [2001].) Zhou [2001] bases pricing on first passage of a jump-diffusion.

B. – Ito's formula

This appendix states Ito's formula, allowing for jumps, and including some background properties of semimartingales. A standard source is Protter [2004]. We first establish some preliminary definitions. We fix a complete probability space $(\Omega, \mathcal{F}, \mathbb{P})$ and a filtration $\{\mathcal{G}_t : t \geq 0\}$ satisfying the *usual conditions*:

- For all t, $\mathcal{G}_t$ contains all of the null sets of $\mathcal{F}$.
- For all t, $\mathcal{G}_t = \cap_{s>t} \mathcal{G}_s$, a property called *right-continuity*.

A function $Z : [0, \infty) \to \mathbb{R}$ is left-continuous if, for all t, we have $Z_t = \lim_{s \uparrow t} Z_s$; the process has left limits if $Z_{t-} = \lim_{s \uparrow t} Z_s$ exists; and finally the process is right-continuous if $Z_t = \lim_{s \downarrow t} Z_s$. The jump ΔZ of Z at time t is $\Delta Z_t = Z_t - Z_{t-}$. The class of processes that are right-continuous with left limits is called RCLL, or sometimes "cadlag," for "continué à, limité à gauche."

Under the usual conditions, we can without loss of generality for our applications assume that a martingale has sample paths that are almost surely right-continuous with left limits. See, for example, Protter [2004], page 8. This is sometimes taken as a defining property of martingales, for example by Jacod and Shiryaev [1987].

LEMMA 1. *Suppose $\mathbb{Q}$ is equivalent to $\mathbb{P}$, with density process ξ. Then an adapted process Y that is right-continuous with left limits is a $\mathbb{Q}$-martingale if and only if ξY is a $\mathbb{P}$-martingale.*

A process X is a finite-variation process if $X = U - V$, where U and V are right-continuous increasing adapted processes with left limits. For example, X is finite-variation if $X_t = \int_0^t \delta_s \, ds$, where δ is an adapted process such that the integral exists. The next lemma is a variant of Ito's formula.

LEMMA 2. *Suppose X is a finite-variation process and $f : \mathbb{R} \to \mathbb{R}$ is continuously differentiable. Then*

$$f(X_t) = f(X_0) + \int_{0+}^{t} f'(X_{s-}) \, dX_s + \sum_{0 < s \leq t} [f(X_s) - f(X_{s-}) - f'(X_{s-})\Delta X_s].$$

Like our next version of Itô's formula, this can be found, for example, in Protter [2004], page 71.

A semimartingale is a process of the form $V + M$, where V is a finite-variation process and M is a local martingale.

LEMMA 3. *Suppose X and Y are semimartingales and at least one of them is a finite-variation process. Let $Z = XY$. Then Z is a semimartingale and*

(B.1) $$dZ_t = X_{t-} \, dY_t + Y_{t-} \, dX_t + \Delta X_t \Delta Y_t.$$

We now extend the last two lemmas. From this point, B denotes a standard Brownian motion in $\mathbb{R}^d$.

LEMMA 4. *Suppose $X = M + A$, where A is a finite-variation process in $\mathbb{R}^d$ and $M_t = \int_0^t \sigma_u \, dB_u$ is in $\mathbb{R}^d$, where B is a standard Brownian motion in $\mathbb{R}^d$ and σ is an $\mathbb{R}^{d \times d}$ progressively-measurable adapted process with $\int_0^t \|\sigma_s\|^2 \, ds < \infty$ almost surely for all t. Suppose $f : \mathbb{R}^d \to \mathbb{R}$ is twice continuously differentiable. Then*

$$f(X_t) = f(X_0) + \int_{0+}^{t} \nabla f(X_{s-}) \, dX_s + \frac{1}{2} \sum_{i,j} \int_0^t \partial_{ij}^2 f(X_s) \, (\sigma_s \sigma_s^\top)_{ij} \, ds$$

$$+ \sum_{0 < s \leq t} [f(X_s) - f(X_{s-}) - \nabla f(X_{s-})\Delta X_s],$$

where $\Delta X(t) = X(t) - X(t-)$ is the jump of X at t and

$$(\nabla f(x))_i = \frac{\partial f(x)}{\partial x_i}, \quad \partial_{ij}^2 f(x) = \frac{\partial^2 f(x)}{\partial x_i \partial x_j}.$$

LEMMA 5. *Suppose $dX_t = dA_t + \sigma_t \, dB_t$ and $dY_t = dC_t + v_t \, dB_t$, where B is a standard Brownian motion in $\mathbb{R}^d$, and where A and C are finite-variation processes, and σ and v are progressively measurable processes in $\mathbb{R}^d$ such that $\int_0^t \sigma_s \cdot \sigma_s \, ds$ and $\int_0^t v_s \cdot v_s \, ds$ are finite almost surely for all t. Let $Z = XY$. Then Z is a semimartingale and*

(B.2) $$dZ_t = X_{t-} \, dY_t + Y_{t-} \, dX_t + \Delta X_t \Delta Y_t + \sigma_t \cdot v_t \, dt.$$

C. – Foundations of Affine Processes

This appendix, based on Duffie, Filipović, and Schachermayer [2003], characterizes regular affine processes, a class of time-homogeneous Markov processes that has arisen from a large and growing range of useful applications in finance. Given a state space of the form $D = \mathbb{R}_+^m \times \mathbb{R}^n$ for integers $m \geq 0$ and $n \geq 0$, the key "affine" property, to be defined precisely in what follows, is roughly that the characteristic exponent (the logarithm of the characteristic function) of the transition distribution $p_t(x, \cdot)$ of such a process is affine with respect to the initial state $x \in D$. The coefficients defining this affine relationship are the solutions of a family of ordinary differential equations (ODEs) that are the essence of the tractability of regular affine processes. We review these ODEs, "generalized Riccati equations," and state the precise set of admissible parameters for which there exists a unique associated regular affine process.

The class of regular affine processes include the entire class of continuous-state branching processes with immigration (CBI) (for example, Kawazu and Watanabe [1971]), as well as the class of processes of the Ornstein-Uhlenbeck (OU) type (for example, Sato [1999]). Roughly speaking, the regular affine processes with state space $\mathbb{R}_+^m$ are CBI, and those with state space $\mathbb{R}^n$ are of OU type. For any regular affine process $X = (Y, Z)$ in $\mathbb{R}_+^m \times \mathbb{R}^n$, the first component Y is necessarily a CBI process. Any CBI or OU process is infinitely decomposable, as is apparent from the exponential-affine form of the characteristic function of its transition distribution. A regular (to be defined below) Markov process with state space D is infinitely decomposable if and only if it is a regular affine process. Regular affine processes are also semimartingales, a crucial property in most financial applications because the standard model of the financial gain generated by trading a security is a stochastic integral with respect to the underlying price process.

We will restrict our attention to the case of time-homogeneous conservative processes (no killing) throughout. For the case that allows for killing, see Duffie, Filipović, and Schachermayer [2003]. For the case of time-inhomogeneous affine processes, see Filipović [2001].

The remainder of the appendix is organized as follows. In Section C.2, we provide the definition of a regular affine process X (Definitions C.1 and C.3) and the main characterization result of affine processes. Three other equivalent characterizations of regular affine processes are then reviewed: (i) in terms of the generator (Theorem C.5), (ii) in terms of the semimartingale characteristics (Theorem C.8), and (iii) in terms of the infinite decomposability (Theorem C.10). Proofs of these results are found in Duffie, Filipović, and Schachermayer [2003].

C.1. – Basic Notation

For background and notation we refer to Jacod and Schiryaev [1987] and Revuz and Yor [1994]. Let $k \in \mathbb{N}$. For α and β in $\mathbb{C}^k$, we write $\langle \alpha, \beta \rangle :=$ $\alpha_1 \beta_1 + \cdots + \alpha_k \beta_k$ (notice that this is not the scalar product on $\mathbb{C}^k$). We let

Sem^k be the convex cone of symmetric positive semi-definite $k \times k$ matrices.

If U is an open set or the closure of an open set in $\mathbb{C}^k$, we write $\overline{U}$ for the closure, U^0 for the interior, and $\partial U = \overline{U} \setminus U^0$ for the boundary.

We use the following notation for function spaces:

- $C(U)$ is the space of complex-valued continuous functions f on U.
- bU is the Banach space of bounded complex-valued Borel-measurable functions f on U.
- $C_b(U)$ is the Banach space $C(U) \cap bU$.
- $C^k(U)$ is the space of k times differentiable functions f on U^0 such that all partial derivatives of f up to order k belong to $C(U)$.
- $C_c^k(U)$ is the space of $f \in C^k(U)$ with compact support.
- $C^\infty(U) = \bigcap_{k \in \mathbb{N}} C^k(U)$ and $C_c^\infty(U) = \bigcap_{k \in \mathbb{N}} C_c^k(U)$.

C.2. – Definition and Characterization

We consider a conservative time-homogeneous Markov process with state space $D = \mathbb{R}_+^m \times \mathbb{R}^n$ and semigroup (P_t) acting on bD,

$$P_t f(x) = \int_D f(\xi) \, p_t(x, d\xi).$$

According to the product structure of D we shall write $x = (y, z)$ or $\xi = (\eta, \zeta)$ for a point in D. We assume that $d = m + n \in \mathbb{N}$. Here, m or n may be zero. We let $(X, (\mathbb{P}_x)_{x \in D}) = ((Y, Z), (\mathbb{P}_x)_{x \in D})$ denote the canonical realization of (P_t) defined on $(\Omega, \mathcal{F}^0, (\mathcal{F}_t^0))$, where Ω is the set of mappings $\omega : \mathbb{R}_+ \to D$ and $X_t(w) = (Y_t(\omega), Z_t(\omega)) = \omega(t)$. The filtration $(\mathcal{F}_t^0)$ is generated by X, and $\mathcal{F}^0 = vee_{t \in \mathbb{R}_+} \mathcal{F}_t^0$. For every $x \in D$, $\mathbb{P}_x$ is a probability measure on $(\Omega, \mathcal{F}^0)$ such that $\mathbb{P}_x[X_0 = x] = 1$ and the Markov property holds, in that, for all s and t in $\mathbb{R}_+$, and for all $f \in bD$,

(C.1) $\qquad \mathbb{E}_x[f(X_{t+s}) \mid \mathcal{F}_t^0] = P_s f(X_t) = \mathbb{E}_{X_t}[f(X_s)], \qquad \mathbb{P}_x\text{-a.s.,}$

where $\mathbb{E}_x$ denotes the expectation with respect to $\mathbb{P}_x$.

For $u = (v, w) \in \mathbb{C}^m \times \mathbb{C}^n$, we write $\breve{u} := (-v, iw) \in \mathbb{C}^m \times \mathbb{C}^n$ and let the function $f_u \in C(D)$ be given by

$$f_u(x) := e^{\langle \breve{u}, x \rangle} = e^{-\langle v, y \rangle + i \langle w, z \rangle}, \qquad x = (y, z) \in D.$$

Notice that $f_u \in C_b(D)$ if and only if $u \in \mathcal{U} := \mathbb{C}_+^m \times \mathbb{R}^n$. By dominated convergence, $P_t f_u(x)$ is continuous in $u \in \mathcal{U}$, for every $(t, x) \in \mathbb{R}_+ \times D$.

Observe that, with a slight abuse of notation,

$$\partial \mathcal{U} \ni u \mapsto P_t f_u(x)$$

is the characteristic function of the measure $p_t(x, \cdot)$, that is, the characteristic function of X_t with respect to $\mathbb{P}_x$.

DEFINITION C.1. The Markov process $(X, (\mathbb{P}_x)_{x \in D})$, and (P_t), is called *affine* if, for every $t \in \mathbb{R}_+$, the characteristic exponent of $p_t(x, \cdot)$ has affine dependence on x. That is, if for every $(t, u) \in \mathbb{R}_+ \times \partial \mathcal{U}$ there exist $\varphi(t, u) \in \mathbb{C}$ and $\psi(t, u) = (\psi^{\mathcal{Y}}(t, u), \psi^{\mathcal{Z}}(t, u)) \in \mathbb{C}^m \times \mathbb{C}^n$ such that

$$(C.2) \qquad P_t f_u(x) = e^{-\varphi(t,u) + \langle \check{\psi}(t,u), x \rangle}$$

$$(C.3) \qquad = e^{-\varphi(t,u) - \langle \psi^{\mathcal{Y}}(t,u), y \rangle + i \langle \psi^{\mathcal{Z}}(t,u), z \rangle}, \quad x = (y, z) \in D.$$

Because $P_t f_u$ is in bD for all $(t, u) \in \mathbb{R}_+ \times \mathcal{U}$, we infer from (C.2) that, *a fortiori* $\varphi(t, u) \in \mathbb{C}_+$ and $\psi(t, u) = (\psi^{\mathcal{Y}}(t, u), \psi^{\mathcal{Z}}(t, u)) \in \mathcal{U}$ for all $(t, u) \in \mathbb{R}_+ \times \partial \mathcal{U}$.

We note that $\psi(t, u)$ is uniquely specified by (C.2), but that $\operatorname{Im} \varphi(t, u)$ is determined only up to multiples of 2π. Nevertheless, by definition we have $P_t f_u(0) \neq 0$ for all $(t, u) \in \mathbb{R}_+ \times \partial \mathcal{U}$. Since $\partial \mathcal{U}$ is simply connected, $P_t f_u(0)$ admits a unique representation of the form (C.2) *–and we shall use the symbol $\varphi(t, u)$ in this sense from now on –* such that $\varphi(t, \cdot)$ is continuous on $\partial \mathcal{U}$ and $\varphi(t, 0) = 0$.

DEFINITION C.2. The Markov process $(X, (\mathbb{P}_x)_{x \in D})$, and (P_t), is *stochastically continuous* if $p_s(x, \cdot) \to p_t(x, \cdot)$ weakly on D, for $s \to t$, for every $(t, x) \in \mathbb{R}_+ \times D$.

If $(X, (\mathbb{P}_x)_{x \in D})$ is affine then, by the continuity theorem of Lévy, $(X, (\mathbb{P}_x)_{x \in D})$ is stochastically continuous if and only if $\varphi(t, u)$ and $\psi(t, u)$ from (C.2) are continuous in $t \in \mathbb{R}_+$, for every $u \in \partial \mathcal{U}$.

DEFINITION C.3 The Markov process $(X, (\mathbb{P}_x)_{x \in D})$, and (P_t), is called *regular* if it is stochastically continuous and the right-hand derivative

$$\widetilde{A} f_u(x) := \partial_t^+ P_t f_u(x)|_{t=0}$$

exists, for all $(x, u) \in D \times \mathcal{U}$, and is continuous at $u = 0$, for all $x \in D$.

We call $(X, (\mathbb{P}_x)_{x \in D})$ and (P_t), *regular affine* if both regular and affine.

If there is no ambiguity, we shall write indifferently X or (Y, Z) for the Markov process $(X, (\mathbb{P}_x)_{x \in D})$, and say that X is *affine, stochastically continuous, regular*, or *regular affine* if $(X, (\mathbb{P}_x)_{x \in D})$ shares the respective property.

In order to state the main characterization results, we require certain notation and terminology. We denote by $\{e_1, \ldots, e_d\}$ the standard basis in $\mathbb{R}^d$, and write $\mathcal{I} := \{1, \ldots, m\}$ and $\mathcal{J} := \{m + 1, \ldots, d\}$. We define the continuous truncation function $\chi = (\chi_1, \ldots, \chi_d) : \mathbb{R}^d \to [-1, 1]^d$ by

$$(C.4) \qquad \begin{aligned} \chi_k(\xi) &= 0, \qquad \xi_k = 0, \\ &= \frac{1 \wedge |\xi_k|}{|\xi_k|} \xi_k, \qquad \text{otherwise.} \end{aligned}$$

Let $\alpha = (\alpha_{ij})$ be a $d \times d$-matrix, $\beta = (\beta_1, \ldots, \beta_d)$ a d-tuple and $I, J \subset \{1, \ldots, d\}$. Then we write α^T for the transpose of α, and $\alpha_{IJ} := (\alpha_{ij})_{i \in I, j \in J}$ and $\beta_I := (\beta_i)_{i \in I}$. Examples are $\chi_I(\xi) = (\chi_k(\xi))_{k \in I}$ or $\nabla_I := (\partial_{x_k})_{k \in I}$. Accordingly, we have $\psi^{\mathcal{Y}}(t, u) = \psi_{\mathcal{I}}(t, u)$ and $\psi^{\mathcal{Z}}(t, u) = \psi_{\mathcal{J}}(t, u)$ (since these mappings play a distinguished role we introduced the former, "coordinate-free" notation). We also write $\mathbf{1} := (1, \ldots, 1)$ without specifying the dimension whenever there is no ambiguity. For $i \in \mathcal{I}$ we define $\mathcal{I}(i) := \mathcal{I} \setminus \{i\}$ and $\mathcal{J}(i) := \{i\} \cup \mathcal{J}$, and let $\mathrm{Id}(i)$ denote the $m \times m$-matrix given by $\mathrm{Id}(i)_{kl} = \delta_{ik}\delta_{kl}$, where δ_{kl} is the Kronecker Delta (δ_{kl} equals 1 if $k = l$ and 0 otherwise).

DEFINITION C.4 The parameters $(a, \alpha, b, \beta, m, \mu)$ are called *admissible* if

- $a \in \mathrm{Sem}^d$ with $a_{\mathcal{II}} = 0$ (hence $a_{\mathcal{IJ}} = 0$ and $a_{\mathcal{JI}} = 0$).
- $\alpha = (\alpha_1, \ldots, \alpha_m)$ with $\alpha_i \in \mathrm{Sem}^d$ and $\alpha_{i,\mathcal{II}} = \alpha_{i,ii}\mathrm{Id}(i)$, for all $i \in \mathcal{I}$.
- $b \in D$.
- $\beta \in \mathbb{R}^{d \times d}$ such that $\beta_{\mathcal{IJ}} = 0$ and $\beta_{i\mathcal{I}(i)} \in \mathbb{R}_+^{m-1}$, for all $i \in \mathcal{I}$. (Hence, $\beta_{\mathcal{II}}$ has nonnegative off-diagonal elements).
- m is a Borel measure on $D \setminus \{0\}$ satisfying

$$\int_{D \setminus \{0\}} \left(\langle \chi_{\mathcal{I}}(\xi), \mathbf{1} \rangle + \|\chi_{\mathcal{J}}(\xi)\|^2 \right) m(d\xi) < \infty.$$

- $\mu = (\mu_1, \ldots, \mu_m)$, where every μ_i is a Borel measure on $D \setminus \{0\}$ satisfying $\int_{D \setminus \{0\}} \left(\langle \chi_{\mathcal{I}(i)}(\xi), \mathbf{1} \rangle + \|\chi_{\mathcal{J}(i)}(\xi)\|^2 \right) \mu_i(d\xi) < \infty$.

Now we have the main characterization results from Duffie, Filipović, and Schachermayer [2003]. First, we state an analytic characterization result for regular affine processes.

THEOREM C.5. *Suppose X is regular affine. Then X is a Feller process. Let A be its infinitesimal generator. Then $C_c^\infty(D)$ is a core of A, $C_c^2(D) \subset \mathcal{D}(A)$, and there exist admissible parameters $(a, \alpha, b, \beta, m, \mu)$ such that, for $f \in C_c^2(D)$,*

$$Af(x) = \sum_{k,l=1}^d \left(a_{kl} + \langle \alpha_{\mathcal{I},kl}, y \rangle \right) \frac{\partial^2 f(x)}{\partial x_k \partial x_l} + \langle b + \beta x, \nabla f(x) \rangle$$

(C.5)
$$+ \int_{D \setminus \{0\}} Gf_0(x, \xi) \, m(d\xi)$$

$$+ \sum_{i=1}^m \int_{D \setminus \{0\}} Gf_i(x, \xi) \, y_i \mu_i(d\xi),$$

where

$$Gf_0(x, \xi) = f(x + \xi) - f(x) - \langle \nabla_{\mathcal{J}} f(x), \chi_{\mathcal{J}}(\xi) \rangle$$
$$Gf_i(x, \xi) = f(x + \xi) - f(x) - \langle \nabla_{\mathcal{J}(i)} f(x), \chi_{\mathcal{J}(i)}(\xi) \rangle .$$

Moreover, (C.2) holds for all $(t, u) \in \mathbb{R}_+ \times \mathcal{U}$ *where* $\varphi(t, u)$ *and* $\psi(t, u)$ *solve the generalized Riccati equations,*

(C.6)
$$\varphi(t, u) = \int_0^t F(\psi(s, u))\, ds$$

(C.7)
$$\partial_t \psi^{\mathcal{Y}}(t, u) = R^{\mathcal{Y}}\big(\psi^{\mathcal{Y}}(t, u), e^{\beta^{\mathcal{Z}} t} w\big), \quad \psi^{\mathcal{Y}}(0, u) = v$$

(C.8)
$$\psi^{\mathcal{Z}}(t, u) = e^{\beta^{\mathcal{Z}} t} w$$

with

(C.9)
$$F(u) = -\langle a\breve{u}, \breve{u}\rangle - \langle b, \breve{u}\rangle$$
$$- \int_{D \setminus \{0\}} \big(e^{\langle \breve{u}, \xi\rangle} - 1 - \langle \breve{u}_{\mathcal{J}}, \chi_{\mathcal{J}}(\xi)\rangle\big)\, m(d\xi),$$

(C.10)
$$R_i^{\mathcal{Y}}(u) = -\langle \alpha_i \breve{u}, \breve{u}\rangle - \langle \beta_i^{\mathcal{Y}}, \breve{u}\rangle$$
$$- \int_{D \setminus \{0\}} \big(e^{\langle \breve{u}, \xi\rangle} - 1 - \langle \breve{u}_{\mathcal{J}(i)}, \chi_{\mathcal{J}(i)}(\xi)\rangle\big)\mu_i(d\xi),$$

for $i \in \mathcal{I}$, *and*

(C.11)
$$\beta_i^{\mathcal{Y}} = (\beta^T)_{i\{1,\ldots,d\}} \in \mathbb{R}^d, \quad i \in \mathcal{I},$$

(C.12)
$$\beta^{\mathcal{Z}} = (\beta^T)_{\mathcal{J}\mathcal{J}} \in \mathbb{R}^{n \times n}.$$

Conversely, let $(a, \alpha, b, \beta, m, \mu)$ *be admissible parameters. Then there exists a unique, regular affine semigroup* (P_t) *with infinitesimal generator (C.5), and (C.2) holds for all* $(t, u) \in \mathbb{R}_+ \times \mathcal{U}$ *where* $\varphi(t, u)$ *and* $\psi(t, u)$ *are given by (C.6)-(C.8).*

Equation (C.8) states that $\psi^{\mathcal{Z}}(t, u)$ depends only on (t, w). Hence, for $w = 0$, we infer from (C.2) that the characteristic function of Y_t with respect to $\mathbb{P}_x$,

$$P_t f_{(v,0)}(x) = \int_D e^{-\langle v, \eta\rangle}\, p_t(x, d\xi) = e^{-\varphi(t,v,0) - \langle \psi^{\mathcal{Y}}(t,v,0), y\rangle}, \quad v \in i\mathbb{R}^m,$$

depends only on y. We obtain the following

COROLLARY C.6. *Let* $X = (Y, Z)$ *be regular affine. Then* $(Y, (\mathbb{P}_{(y,z)})_{y \in \mathbb{R}_+^m})$ *is a regular affine Markov process with state space* $\mathbb{R}_+^m$, *and does not depend on* $z \in \mathbb{R}^n$.

Theorem C.5 generalizes and unifies two well studied classes of stochastic processes. For the notion of a *CBI process* we refer to Watanabe [1969], Kawazu and Watanabe [1971] and Shiga and Watanabe [1973]. For the notion of an *OU type process* see (Sato 1999, Definition 17.2).

COROLLARY C.7. *Let* $X = (Y, Z)$ *be regular affine. Then* $(Y, (\mathbb{P}_{(y,z)})_{y \in \mathbb{R}_+^m})$ *is a CBI process, for every* $z \in \mathbb{R}^n$. *If* $m = 0$, *then* X *is an OU-type process. Conversely, every CBI and OU type process is a regular affine Markov process.*

Motivated by Theorem C.5, we give in this paragraph a summary of some classical results for Feller processes. For proofs, we refer to Revuz and Yor [1994], Chapter III.2. Let X be regular affine and hence, by Theorem C.5, a Feller process. Since we deal with an entire family of probability measures, $(\mathbb{P}_x)_{x \in D}$, we use the convention that "a.s." means "$\mathbb{P}_x$-a.s. for all $x \in D$". Then X admits a cadlag modification, and from now on we shall only consider cadlag versions of a regular affine process X, still denoted by X.

We write $\mathcal{F}^{(x)}$ for the completion of $\mathcal{F}^0$ with respect to $\mathbb{P}_x$ and $(\mathcal{F}_t^{(x)})$ for the filtration obtained by adding to each $\mathcal{F}_t^0$ all $\mathbb{P}_x$-nullsets in $\mathcal{F}^{(x)}$. Define

$$\mathcal{F}_t := \bigcap_{x \in D} \mathcal{F}_t^{(x)}, \quad \mathcal{F} := \bigcap_{x \in D} \mathcal{F}^{(x)}.$$

Then the filtrations $(\mathcal{F}_t^{(x)})$ and $(\mathcal{F}_t)$ are right-continuous, and X is still a Markov process with respect to $(\mathcal{F}_t)$. That is, (C.1) holds for $\mathcal{F}_t^0$ replaced by $\mathcal{F}_t$, for all $x \in D$.

By convention, we call X a *semimartingale* if X is a semimartingale on $(\Omega, \mathcal{F}, (\mathcal{F}_t), \mathbb{P}_x)$, for every $x \in D$. For the definition of the *characteristics* of a semimartingale with refer to (Jacod and Shiryaev 1987, Section II.2). We emphasize that the characteristics below are associated with the truncation function χ, defined in (C.4).

Let X' be a D-valued stochastic process defined on some probability space $(\Omega', \mathcal{F}', \mathbb{P}')$. Then $\mathbb{P}' \circ X'^{-1}$ denotes the law of X', that is, the image of $\mathbb{P}'$ by the mapping $\omega' \mapsto X'_{(\cdot)}(\omega') : (\Omega', \mathcal{F}') \to (\Omega, \mathcal{F}^0)$.

The following is a characterization result for regular affine processes in the class of semimartingales.

THEOREM C.8. *Let X be regular affine and $(a, \alpha, b, \beta, m, \mu)$ the related admissible parameters. Then X is a semimartingale and admits the characteristics (B, C, v), where*

(C.13)
$$B_t = \int_0^t (\tilde{b} + \tilde{\beta} X_s)\, ds$$

(C.14)
$$C_t = 2 \int_0^t \left(a + \sum_{i=1}^m \alpha_i Y_s^i \right) ds$$

(C.15)
$$v(dt, d\xi) = \left(m(d\xi) + \sum_{i=1}^m Y_t^i \mu_i(d\xi) \right) dt$$

for every $\mathbb{P}_x$, where $\tilde{b} \in D$ and $\tilde{\beta} \in \mathbb{R}^{d \times d}$ are given by

(C.16)
$$\tilde{b} = b + \int_{D \setminus \{0\}} (\chi_{\mathcal{I}}(\xi), 0)\, m(d\xi),$$

(C.17)
$$\tilde{\beta}_{kl} = \beta_{kl} + (1 - \delta_{kl}) \int_{D \setminus \{0\}} \chi_k(\xi)\, \mu_l(d\xi), \quad \text{if } l \in \mathcal{I},$$

(C.18)
$$= \beta_{kl}, \quad \text{if } l \in \mathcal{J}, \quad \text{for } 1 \le k \le d.$$

Moreover, let $X' = (Y', Z')$ *be a D-valued semimartingale defined on some filtered probability space* $(\Omega', \mathcal{F}', (\mathcal{F}'_t), \mathbb{P}')$ *with* $\mathbb{P}'[X'_0 = x] = 1$. *Suppose that* X' *admits the characteristics* (B', C', ν'), *given by* (C.13)-(C.15) *where* X *is replaced by* X'. *Then* $\mathbb{P}' \circ X'^{-1} = \mathbb{P}_x$.

A third way of characterizing regular affine processes, generalizes Shiga and Watanabe [1973], as follows. Let $\mathbb{P}$ and $\mathbb{Q}$ be two probability measures on $(\Omega, \mathcal{F}^0)$. We write $\mathbb{P} * \mathbb{Q}$ for the image of $\mathbb{P} \times \mathbb{Q}$ by the measurable mapping $(\omega, \omega') \mapsto \omega + \omega' : (\Omega \times \Omega, \mathcal{F}^0 \otimes \mathcal{F}^0) \to (\Omega, \mathcal{F}^0)$. Let $\mathcal{P}_{RM}$ be the set of all families $(\mathbb{P}_x)_{x \in D}$ of probability measures on $(\Omega, \mathcal{F}^0)$ such that $(X, (\mathbb{P}_x)_{x \in D})$ is a regular Markov process with $\mathbb{P}_x[X_0 = x] = 1$, for all $x \in D$.

DEFINITION C.9. We call $(\mathbb{P}_x)_{x \in D}$ *infinitely decomposable* if, for every $k \in \mathbb{N}$, there exists $(\mathbb{P}_x^{(k)})_{x \in D} \in \mathcal{P}_{RM}$ such that

$$(C.19) \qquad \mathbb{P}_{x^{(1)} + \cdots + x^{(k)}} = \mathbb{P}_{x^{(1)}}^{(k)} * \cdots * \mathbb{P}_{x^{(k)}}^{(k)}, \qquad \text{for all } x^{(1)}, \dots, x^{(k)} \in D.$$

THEOREM C.10. *The Markov process* $(X, (\mathbb{P}_x)_{x \in D})$ *is regular affine if and only if* $(\mathbb{P}_x)_{x \in D}$ *is infinitely decomposable.*

Without going much into detail, we remark that in (C.5) we can distinguish the three "building blocks" of any jump-diffusion process, the *diffusion matrix* $A(x) = a + y_1 \alpha_1 + \cdots + y_m \alpha_m$, the *drift* $B(x) = b + \beta x$, and the *Lévy measure* (the compensator of the jumps) $M(x, d\xi) = m(d\xi) + y_1 \mu_1(d\xi) + \cdots + y_m \mu_m(d\xi)$. An informal definition of an affine process could consist of the requirement that $A(x)$, $B(x)$, and $M(x, d\xi)$ have affine dependence on x. (See, for example, Duffie, Pan, and Singleton [2000].) The particular kind of this affine dependence in the present setup is implied in part by the geometry of the state space D.

D. – Toolbox for Affine Processes

This appendix contains some "tools" for affine processes that are implications of the preceding basic results.

D.1. – Statistical Estimation of Affine Models

For a given (regular) affine process X with state space $D \subset \mathbb{R}^d$, we reconsider the conditional characteristic function φ of X_T given X_t, defined from (3.1) by

$$(D.1) \qquad \Phi(u, X_t, t, T) = E\big(e^{iu \cdot X_T} | X_t\big),$$

for real u in $\mathbb{R}^d$. Because knowledge of Φ is equivalent to knowledge of the joint conditional transition distribution function of X, this result is useful in estimation and all other applications involving the transition densities of affine processes.

For instance, Singleton [2001] exploits knowledge of Φ to derive maximum likelihood estimators for the coefficients of an affine process, based on the conditional density $f(\cdot \mid X_t)$ of X_{t+1} given X_t, obtained by Fourier inversion of φ as

(D.2) $$f(X_{t+1} \mid X_t) = \frac{1}{(2\pi)^N} \int_{\mathbb{R}^N} e^{-iu \cdot X_{t+1}} \Phi(u, X_t, t, t+1) \, du.$$

Das [1998] exploits (D.2) for special case of an affine process to compute method-of-moments estimators of a model of interest rates.

Method-of-moments estimators can also be constructed directly in terms of the conditional characteristic function. From the definition of Φ,

(D.3) $$E\left[e^{iu \cdot X_{t+1}} - \Phi(u, X_t, t, t+1) \mid X_t\right] = 0,$$

so any measurable function of X_t is orthogonal to the innovation $(e^{iu \cdot X_{t+1}} - \Phi(u, X_t, t, t+1))$. Singleton [2001] uses this fact, together with the known functional form of Φ, to construct generalized method-of-moments estimators of the parameters governing affine processes and, more generally, the parameters of asset pricing models in which the state process is affine. These estimators are computationally tractable and, in some cases, achieve the same asymptotic efficiency as the maximum likelihood estimator. Jiang and Knight [2002] and Chacko and Viceira [2003] propose related, characteristic-function-based estimators of the stochastic volatility model of asset returns in which the instantaneous variance process is a Feller diffusion.

D.2. – Laplace Transforms and Moments

First, for a one-dimensional affine process, we consider the Laplace transform $\varphi(\cdot)$, whenever well-defined at some number u by

(D.4) $$\varphi(u) = E_t\left(e^{-uX(s)}\right).$$

We sometimes call $\varphi(\cdot)$ the moment-generating function of $X(s)$, for it has the convenient property that, if its successive derivatives

$$\varphi'(0), \varphi''(0), \varphi'''(0), \ldots, \varphi^{(m)}(0)$$

up to some order m are well defined, then they provide us with the respective moments

$$\varphi^{(k)}(0) = E_t[X(s)^k].$$

From (3.3), we know that $\varphi(u) = e^{\alpha(t) + \beta(t)X(t)}$, for coefficients $\alpha(t)$ and $\beta(t)$ obtained from the generalized Riccati equation for X. We can calculate the dependence of $\alpha(t)$ and $\beta(t)$ on the boundary condition $\beta(s) = u$, writing

$\alpha(t, u)$ and $\beta(t, u)$ to show this dependence explicitly. Then we have, by the chain rule for differentiation,

$$\varphi'(0) = e^{\alpha(t,0)+\beta(t,0)X(0)}[\alpha_u(t, 0) + \beta_u(t, 0)X(0)],$$

where β_u denotes the partial derivative of β with respect to its boundary condition u. Successively higher-order derivatives can be computed by repeated differentiation. Pan [2002] provides an efficient recursive algorithm for higher-order moments, even in certain multivariate cases.

For the multivariate case, the transform at $u \in \mathbb{R}^d$ is defined by

(D.5) $$\varphi(u) = E\left(e^{-u \cdot X(t)}\right) = e^{\alpha(t,u)+\beta(t,u) \cdot X(0)},$$

and provides covariance and other cross-moments, again by differentiation.

Having an explicit term structure of such moments as variances, covariances, skewness, kurtosis, and so on, as the time horizon s varies, allows one to analytically calibrate models to data, or to formulate models in light of empirical regularities, as shown by Das and Sundaram [1999]. For example, method-of-moments statistical estimation in a time-series setting can also be based on the conditional moment-generating function (D.4).

Gregory and Laurent [2003] extend the transform to a generating function designed to simplify the calculation of the distribution function of the number of defaults from a portfolio of defaultable bonds.

D.3. – Inversion of the Transform

The probability distribution of a random variable Z can be recovered from its characteristic function $\Psi(\cdot)$ by the Lévy inversion formula, according to which (under technical regularity conditions),

(D.6) $$\mathbb{P}(Z \leq z) = \frac{\Psi(0)}{2} - \frac{1}{\pi} \int_0^\infty \frac{\text{Im}\left[\Psi(u)e^{-iuy}\right]}{u} du,$$

where $\text{Im}(c)$ denotes the imaginary part of any complex number c. It is sufficient, for example, that $\int |\Psi(u)| du < +\infty$.

The integral in (D.6) is typically calculated by a numerical method such as quadrature, which is rapid.

For affine processes, in a few cases, such as the Ornstein-Uhlenbeck (Gaussian) model and the Feller diffusion (non-central χ^2), the probability transition distribution is known explicitly.

D.4. – Solution for The Basic Affine Model

This section summarizes some results from Duffie and Gârleanu [2001] for the solutions β and α for the basic affine model X of (3.11).

These generalized Riccati equations reduce in this special case to the form

(D.7)
$$\frac{d\beta(t)}{dt} = n\beta(t) + \frac{1}{2}p\beta(t)^2 + q$$

(D.8)
$$\frac{d\alpha(t)}{dt} = m\beta(t) + \ell\frac{\bar{\mu}\beta(t)}{1 - \bar{\mu}\beta(t)},$$

for some constant coefficients n, p, q, m, ℓ, and $\bar{\mu}$, with boundary conditions $\alpha(s) = u$ and $\beta(s) = v$. We may take complex boundary conditions u and v, for example in order to recover the characteristic function $\varphi(\theta) = E_x(e^{i\theta X(s)})$, by taking $q = 0$, $u = 0$, and $v = i\theta$ for real θ.

More generally, the expectation

(D.9)
$$E_x\left[e^{\int_0^s qX(z)\,dz + v + uX(s)}\right] = e^{\alpha(s) + \beta(s)x},$$

has explicit solutions for $\alpha(s)$ and $\beta(s)$ given below. For example, the discount $E(e^{-\int_0^t X(s)\,ds}) = e^{\alpha(t) + \beta(t)X(0)}$ is obtained by taking $u = v = 0$, $n = -\kappa$, $p = \sigma^2$, $q = -1$, and $m = \kappa\theta$. In general, solutions are given by

(D.10)
$$\beta(s) = \frac{1 + a_1 e^{b_1 s}}{c_1 + d_1 e^{b_1 s}}$$

$$\alpha(s) = v + \frac{m(a_1 c_1 - d_1)}{b_1 c_1 d_1} \log \frac{c_1 + d_1 e^{b_1 s}}{c_1 + d_1} + \frac{m}{c_1}s$$

(D.11)
$$+ \frac{\ell(a_2 c_2 - d_2)}{b_2 c_2 d_2} \log \frac{c_2 + d_2 e^{b_2 s}}{c_2 + d_2} + \left(\frac{\ell}{c_2} - \ell\right)s,$$

where

$$c_1 = \frac{-n + \sqrt{n^2 - 2pq}}{2q}$$

$$d_1 = (1 - c_1 u)\frac{n + pu + \sqrt{(n + pu)^2 - p(pu^2 + 2nu + 2q)}}{2nu + pu^2 + 2q}$$

$$a_1 = (d_1 + c_1)u - 1$$

$$b_1 = \frac{d_1(n + 2qc_1) + a_1(nc_1 + p)}{a_1 c_1 - d_1}$$

$$a_2 = \frac{d_1}{c_1}$$

$$b_2 = b_1$$

$$c_2 = 1 - \frac{\bar{\mu}}{c_1}$$

$$d_2 = \frac{d_1 - \bar{\mu}a_1}{c_1}.$$

E. – Doubly Stochastic Counting Processes

This appendix, based on Appendix I of Duffie [2001], reviews the basic theory of intensity-based models of counting processes. Brémaud [1981] is a standard source. We emphasize the doubly-stochastic setting, because of its tractability.

All properties below are with respect to a probability space $(\Omega, \mathcal{F}, \mathbb{P})$ and a given filtration $\{\mathcal{G}_t : t \geq 0\}$ satisfying the usual conditions (Appendix D) unless otherwise indicated. We sometimes use the usual shorthand, as in "$(\mathcal{F}_t)$-adapted," to specify a property with respect to some alternative filtration $\{\mathcal{F}_t : t \geq 0\}$.

A process Y is predictable if $Y : \Omega \times [0, \infty) \to \mathbb{R}$ is measurable with respect to the σ-algebra on $\Omega \times [0, \infty)$ generated by the set of all left-continuous adapted processes. The idea is that one can "foretell" Y_t based on all of the information available up to, but not including, time t. Of course, any left-continuous adapted process is predictable, as is, in particular, any continuous process.

A counting process N, also known as a point process, is defined via an increasing sequence $\{T_0, T_1, \ldots\}$ of random variables valued in $[0, \infty]$, with $T_0 = 0$ and with $T_n < T_{n+1}$ whenever $T_n < \infty$, according to

(E.1) $$N_t = n, \quad t \in [T_n, T_{n+1}),$$

where we define $N_t = +\infty$ if $t \geq \lim_n T_n$. We may treat T_n as the n-th jump time of N, and N_t as the number of jumps that have occurred up to and including time t. The counting process is nonexplosive if $\lim T_n = +\infty$ almost surely.

Definitions of "intensity" vary slightly from place to place. One may refer to Section II.3 of Brémaud [1981], in particular Theorems T8 and T9, to compare other definitions of intensity with the following. Let λ be a nonnegative predictable process such that, for all t, we have $\int_0^t \lambda_s \, ds < \infty$ almost surely. Then a nonexplosive adapted counting process N has λ as its intensity if $\{N_t - \int_0^t \lambda_s \, ds : t \geq 0\}$ is a local martingale.

From Brémaud's Theorem T12, without an important loss of generality for our purposes, we can require an intensity to be predictable, as above, and we can treat an intensity as essentially unique, in that: If λ and $\tilde{\lambda}$ are both intensities for N, as defined above, then

(E.2) $$\int_0^\infty |\lambda_s - \tilde{\lambda}_s| \lambda_s \, ds = 0 \quad \text{a.s.}$$

We note that if λ is strictly positive, then (E.2) implies that $\lambda = \tilde{\lambda}$ almost everywhere.

We can get rid of the annoying "localness" of the above local-martingale characterization of intensity under the following technical condition, which can be verified from Theorems T8 and T9 of Brémaud [1981].

PROPOSITION 1. *Suppose N is an adapted counting process and λ is a nonnegative predictable process such that, for all t, $E(\int_0^t \lambda_s \, ds) < \infty$. Then the following are equivalent:*

(i) *N is nonexplosive and λ is the intensity of N.*

(ii) *$\{N_t - \int_0^t \lambda_s \, ds : t \geq 0\}$ is a martingale.*

PROPOSITION 2. *Suppose N is a nonexplosive adapted counting process with intensity λ, with $\int_0^t \lambda_s \, ds < \infty$ almost surely for all t. Let M be defined by $M_t = N_t - \int_0^t \lambda_s \, ds$. Then, for any predictable process H such that $\int_0^t |H_s| \lambda_s \, ds$ is finite almost surely for all t, a local martingale Y is well defined by*

$$Y_t = \int_0^t H_s \, dM_s = \int_0^t H_s \, dN_s - \int_0^t H_s \lambda_s \, ds.$$

If, moreover, $E\left[\int_0^t |H_s| \lambda_s \, ds\right] < \infty$, then Y is a martingale.

In order to define a Poisson process, we first recall that a random variable K with outcomes $\{0, 1, 2, \ldots\}$ has the Poisson distribution with parameter β if

$$\mathbb{P}(K = k) = e^{-\beta} \frac{\beta^k}{k!},$$

noting that $0! = 1$. A Poisson process is an adapted nonexplosive counting process N with deterministic intensity λ such that $\int_0^t \lambda_s \, ds$ is finite almost surely for all t, with the property that, for all t and $s > t$, conditional on $\mathcal{G}_t$, the random variable $N_s - N_t$ has the Poisson distribution with parameter $\int_t^s \lambda_u \, du$. (See Brémaud [1981], page 22.)

We recall that $\{\mathcal{G}_t : t \in [0, T]\}$ is the augmented filtration of a process Y valued in some Euclidean space if, for all t, $\mathcal{G}_t$ is the completion of $\sigma(\{Y_s : 0 \leq s \leq t\})$.

Suppose N is a nonexplosive counting process with intensity λ, and $\{\mathcal{F}_t : t \geq 0\}$ is a filtration satisfying the usual conditions, with $\mathcal{F}_t \subset \mathcal{G}_t$. We say that N is doubly stochastic, driven by $\{\mathcal{F}_t : t \geq 0\}$, if λ is $(\mathcal{F}_t)$-predictable and if, for all t and $s > t$, conditional on the σ-algebra $\mathcal{G}_t \vee \mathcal{F}_s$ generated by $\mathcal{G}_t \cup \mathcal{F}_s$, $N_s - N_t$ has the Poisson distribution with parameter $\int_t^s \lambda_u \, du$. For conditions under which a filtration $\{\mathcal{F}_t : t \geq 0\}$ generated by a Markov process X satisfies the usual conditions, see Chung [1982], Theorem 4, page 61. We can extend to the multi-type counting process $N = (N^{(1)}, \ldots, N^{(k)})$ that is doubly stochastic driven by $\{\mathcal{F}_t : t \geq 0\}$ with intensity $\lambda = (\lambda_1, \ldots, \lambda_k)$. The same definition applies to each coordinate counting process $N^{(i)}$, and moreover, "conditional on" the driving filtration $\{\mathcal{F}_t : t \geq 0\}$ the coordinate processes $N^{(1)}, \ldots, N^{(k)}$ are independent. That is, conditional on the σ-algebra $\mathcal{G}_t \vee \mathcal{F}_s$ generated by $\mathcal{G}_t \cup \mathcal{F}_s$, $\{\{N_u^{(i)} - N_t^{(i)} : t \leq u \leq s\} : 1 \leq i \leq k\}$ are independent.

It is to be emphasized that the filtration $\{\mathcal{G}_t : t \geq 0\}$ has been fixed in advance for purposes of the above definitions. In applications involving doubly stochastic processes, it is often the case that one constructs the underlying filtration $\{\mathcal{G}_t : t \geq 0\}$ as follows. First, one has a filtration $\{\mathcal{F}_t : t \geq 0\}$ satisfying the usual conditions, and an $(\mathcal{F}_t)$-predictable process λ such that $\int_0^t \lambda_s \, ds < \infty$ almost surely for all t. We then let $Z_1, Z_2, \ldots$ be independent standard exponential random variables (that is, with $\mathbb{P}(Z_i > z) = e^{-z}$) that, for all t, are independent of $\mathcal{F}_t$. We let $T_0 = 0$ and, for $n \geq 1$, we let T_n be defined recursively by

$$(E.3) \qquad T_n = \inf \left\{ t \geq T_{n-1} : \int_{T_{n-1}}^t \lambda_u \, du = Z_n \right\}.$$

Now, we can define N by (E.1). (This construction can also be used for Monte Carlo simulation of the jump times of N.) Finally, for each t, we let $\mathcal{G}_t$ be the σ-algebra generated by $\mathcal{F}_t$ and $\{N_s : 0 \leq s \leq t\}$. By this construction, relative to the filtration $\{\mathcal{G}_t : t \geq 0\}$, N is a nonexplosive counting process with intensity λ that is doubly stochastic driven by $\{\mathcal{F}_t : t \geq 0\}$. The construction for the multi-type case is the obvious extension by which the set of exponential random variables triggering arrivals for the respective types are independent. In examples, $\{\mathcal{F}_t : t \geq 0\}$ is usually the filtration generated by a Markov state process.

Next, we review the martingale representation theorem in this setting, restricting attention to a fixed time interval $[0, T]$. We say that a local martingale $M = (M^{(1)}, \ldots, M^{(k)})$ in $\mathbb{R}^k$ has the martingale representation property if, for any martingale Y, there exist predictable processes $H^{(1)}, \ldots, H^{(k)}$ such that the stochastic integral $\int H^{(i)} \, dM^{(i)}$ is well defined for each i and

$$Y_t = Y_0 + \sum_{i=1}^k \int_0^t H_s^{(i)} \, dM_s^{(i)}, \quad \text{a.s.} \quad t \in [0, T].$$

The following representation result is from Brémaud [1981].

PROPOSITION 3. *Suppose that $N = (N^{(1)}, \ldots, N^{(k)})$, where $N^{(i)}$ is a nonexplosive counting process that has the intensity $\lambda^{(i)}$ relative to the augmented filtration $\{\mathcal{G}_t : t \in [0, T]\}$ of N. Let $M_t^{(i)} = N_t^{(i)} - \int_0^t \lambda_s^{(i)} \, ds$. Then $M = (M^{(1)}, \ldots, M^{(k)})$ has the martingale representation property for $\{\mathcal{G}_t : t \in [0, T]\}$.*

PROPOSITION 4. *For a given probability space, let $N = (N^{(1)}, \ldots, N^{(k)})$, where $N^{(i)}$ is a Poisson process with intensity $\lambda^{(i)}$ relative to the augmented filtration generated by N itself. Let B be a standard Brownian motion in $\mathbb{R}^d$, independent of N, and let $\{\mathcal{G}_t : t \geq 0\}$ be the augmented filtration generated by (B, N). Then $N^{(i)}$ has the $(\mathcal{G}_t)$-intensity $\lambda^{(i)}$, a $(\mathcal{G}_t)$-martingale $M^{(i)}$ is defined by $M_t^{(i)} = N_t^{(i)} - \int_0^t \lambda_s^{(i)} \, ds$, and $(B^{(1)}, \ldots, B^{(d)}, M^{(1)}, \ldots, M^{(k)})$ has the martingale representation property for $\{\mathcal{G}_t : t \geq 0\}$.*

We can also extend this result as follows. We can let $N^{(i)}$ be doubly stochastic driven by the standard filtration of a standard Brownian motion B in $\mathbb{R}^d$. For the augmented filtration generated by (B, N), defining $M^{(i)}$ as the compensated counting process of $N^{(i)}$, $(B^{(1)}, \ldots, B^{(d)}, M^{(1)}, \ldots, M^{(k)})$ has the martingale representation property. For details and technical conditions, see, for example, Kusuoka [1999]. For more on martingale representation, see Jacod and Shiryaev [1987].

Next, we turn to Girsanov's theorem. Suppose N is a nonexplosive counting process with intensity λ, and φ is a strictly positive predictable process such that, for some fixed time horizon T, $\int_0^T \varphi_s \lambda_s \, ds$ is finite almost surely. A local martingale ξ is then well defined by

$$(E.4) \qquad \xi_t = \exp\left(\int_0^t (1 - \varphi_s)\lambda_s \, ds\right) \prod_{\{i:T(i)\leq t\}} \varphi_{T(i)}, \quad t \leq T,$$

where $T(i)$ denotes the i-th jump time of K.

PROPOSITION 5. *Suppose that the local martingale ξ is a martingale. For this, it suffices that λ is bounded and deterministic and that φ is bounded. Then an equivalent probability measure $\mathbb{Q}$ is defined by letting $\frac{d\mathbb{Q}}{d\mathbb{P}} = \xi(T)$. Restricted to the time interval $[0, T]$, under the probability measure $\mathbb{Q}$, N is a nonexplosive counting process with intensity $\lambda\varphi$.*

The proof is essentially the same as that found in Brémaud [1981], page 168, making use of Lemmas 1 and 2 of Appendix B.

By an extra measurability condition, we can specify a change of probability measure associated with a given change of intensity, under which the doubly stochastic property is preserved.

PROPOSITION 6. *Suppose N is doubly stochastic driven by $\{\mathcal{F}_t : t \geq 0\}$ with intensity λ, where $\mathcal{G}_t$ is the completion of $\mathcal{F}_t \vee \sigma(\{N_s : 0 \leq s \leq t\})$. For a fixed time $T > 0$, let φ be an $(\mathcal{F}_t)$-predictable process with $\int_0^T \varphi_t \lambda_t \, dt < \infty$ almost surely. Let ξ be defined by (E.4), and suppose that ξ is a martingale. (For this, it suffices that λ and φ are bounded.) Let $\mathbb{Q}$ be the probability measure with $\frac{d\mathbb{Q}}{d\mathbb{P}} = \xi_T$. Then, restricted to the time interval $[0, T]$, under the probability measure $\mathbb{Q}$ and with respect to the filtration $\{\mathcal{G}_t : 0 \leq t \leq T\}$, N is doubly stochastic driven by $\{\mathcal{F}_t : t \in [0, T]\}$, with intensity $\lambda\varphi$.*

For a proof, we note that, under $\mathbb{Q}$, N is, by Proposition 5, a nonexplosive counting process with $(\mathcal{G}_t)$-intensity $\lambda\varphi$, which is $(\mathcal{F}_t)$-predictable. Further, ξ is a $\mathbb{P}$-martingale with respect to the filtration $\{\mathcal{H}_t : t \in [0, T]\}$ defined by letting $\mathcal{H}_t$ be the completion of $\sigma(\{N_s : s \leq t\}) \vee \mathcal{F}_T$. (To verify the stated sufficient condition for martingality, we can apply the argument of Brémaud [1981], page 168, the doubly stochastic property under $\mathbb{P}$, and the law of iterated expectations.) Now, we can use the characterization (1.8), page 22, of Brémaud [1981] and apply Ito's formula to see that, under $\mathbb{Q}$ with respect to $\{\mathcal{H}_t : t \in [0, T]\}$, the counting process N is doubly stochastic, driven by $\{\mathcal{F}_t : t \geq 0\}$, with intensity $\lambda\varphi$. Finally, the result follows by noting that, whenever $s > t$, we have

$$\mathbb{Q}(N_s - N_t = k \mid \mathcal{G}_t \vee \mathcal{F}_s) = \mathbb{Q}(N_s - N_t = k \mid \sigma(\{N_u : 0 \leq u \leq t\}) \vee \mathcal{F}_s)$$
$$= \mathbb{Q}(N_s - N_t = k \mid \mathcal{H}_t),$$

using the definition of $\mathcal{G}_t$, and the fact that $\mathcal{F}_t \subset \mathcal{F}_s$.

Finally, we calculate, under some regularity conditions, the density and hazard rate of a doubly stochastic stopping time τ with intensity λ, driven by some filtration.

Letting $p(t) = P(\tau > t)$ define the survival function $p : [0, \infty) \to [0, 1]$, the density of the stopping time τ is $\pi(t) = -p'(t)$, if it exists, and the hazard function $h : [0, \infty) \to [0, \infty)$ is defined by

$$h(t) = \frac{\pi(t)}{p(t)} = -\frac{d}{dt} \log p(t),$$

so that we can then write

$$p(t) = e^{-\int_0^t h(u)\,du}.$$

One can similarly define the $\mathcal{G}_t$-conditional density and hazard rate.

Now, because $p(t) = E(e^{-\int_0^t \lambda(u)\,du})$, if differentiation through this expectation is justified, then we would have the natural result that

(E.5)
$$p'(t) = E\left(-e^{-\int_0^t \lambda(u)\,du} \lambda(t)\right),$$

from which $\pi(t)$ and $h(t)$ would be defined as above. Grandell [1976], pages 106-107, has shown that (E.5) is correct provided that:
 i) There is a constant C such that, for all t, $E(\lambda_t^2) < C$.
 ii) For any $\epsilon > 0$ and almost every time t,

$$\lim_{\delta \to 0} P(|\lambda(t + \delta) - \lambda(t)| \geq \epsilon) = 0.$$

These properties are satisfied in many typical models.

F. – Further Reading

The use of intensity-based defaultable bond pricing models was instigated by Artzner and Delbaen [1990], Artzner and Delbaen [1992], Artzner and Delbaen [1995], Lando [1994], Lando [1998], and Jarrow and Turnbull [1995]. For additional work in this vein, see Berndt, Douglas, Duffie, Fergusen, and Schranz [2003], Bielecki and Rutkowski [2002], Cooper and Mello [1991], Cooper and Mello [1992], Das and Sundaram [1999], Das and Tufano [1995], Davydov, Linetsky, and Lotz [1999], Duffie [1998a], Duffie and Huang [1996], Duffie and Singleton [1999], Elliott, Jeanblanc, and Yor [2000], Hull and White [1992], Hull and White [1995], Jarrow and Yu [2001], Jarrow, Lando, and Yu [1999], Jeanblanc and Rutkowski [2000], Madan and Unal [1998], and Nielsen and Ronn [1995].

Monographs devoted to the subject of credit risk modeling include those of Arvanitis and Gregory [2001], Bielecki and Rutkowski [2002], Bluhm, Overbeck, and Wagner [2003], Duffie and Singleton [2002], Lando [2004], and Schönbucher [2003a].

Intensity-based debt pricing models based on stochastic transition among credit ratings were developed by Arvanitis, Gregory, and Laurent [1999], Jarrow, Lando, and Turnbull [1997], Kijima and Komoribayashi [1998], Kijima [1998], and Lando [1998].

Corporate bond pricing under Gaussian interest rates was explored by Décamps and Rochet [1997] and Shimko, Tejima, and van Deventer [1993]. On the impact of illiquidity on defaultable debt prices, see Ericsson and Renault [1999].

Models of, and empirical work on, default correlation include those of Collin-Dufresne Goldstein, and Hugonnier [2002], Das, Duffie, and Kapadia [2004], Davis and Lo [1999], Davis and Lo [2000], Duffie and Gârleanu [2001], Finger [2000], Schönbucher [2003b], Schönbucher and Schubert [2001], and Yu [2002b].

Bibliography

V. Acharya – J. Carpenter (2001), "Corporate Bond Valuation and Hedging with Stochastic Interest Rates and Endogeneous Bankruptcy", Review of Financial Studies 15, 1355-1383.

E. Altman – B. Brady – A. Resti – A. Sironi (2003), "The Link Between Default and Recovery Rates: Theory, Empirical Evidence and Implications", New York University, forthcoming in Journal of Business.

R. Anderson – Y. Pan – S. Sundaresan (1995), "Corporate Bond Yield Spreads and the Term Structure", Working Paper, CORE, Belgium.

R. Anderson – S. Sundaresan (1996), *Design and Valuation of Debt Contracts*, Review of Financial Studies **9**, 37-68.

P. Artzner – F. Delbaen (1990), *"Finem Lauda" or the Risk of Swaps*, Insurance: Mathematics and Economics **9**, 295-303.

P. Artzner – F. Delbaen (1992), *Credit Risk and Prepayment Option*, ASTIN Bulletin **22**, 81-96.

P. Artzner – F. Delbaen (1995), *Default Risk and Incomplete Insurance Markets*, Mathematical Finance **5**, 187-195.

A. Arvanitis – J. Gregory (2001), "Credit", Risk Books, London.

A. Arvantis – J. Gregory – J.-P. Laurent (1999), *Building Models for Credit Spreads*, Journal of Derivatives **6** (3), 27-43.

G. Bakshi – C. Cao – Z. Chen (1997), *Empirical Performance of Alternative Option Pricing Models*, Journal of Finance **52**, 2003-2049.

G. Bakshi – D. Madan (2000), *Spanning and Derivative Security Valuation*, Journal of Financial Economics **55**, 205-238.

G. Bakshi – D. Madan – F. Zhang (2001), "Investigating the Sources of Default Risk: Lessons from Empirically Evaluating Credit Risk Models", Working Paper, University of Maryland.

D. Bates (1996), *Jumps and Stochastic Volatility: Exchange Rate Processes Implicit in Deutsche Mark Option*, Review of Financial Studies **9**, 69-107.

D. Bates (1997), *Post-87' Crash Fears in S-and-P 500 Futures Options*, Journal of Econometrics **94**, 181-238.

A. Berndt (2002), "Estimating the Term Structure of Credit Spreads: Callable Corporate Debt", Working Paper, Department of Industrial Engineering and Operations Research, Cornell University.

A. Berndt – R. Douglas – D. Duffie – M. Fergusen – D. Schranz (2003), "Default Risk Premia from Default Swap Rates and EDFs", Working Paper, Graduate School of Business, Stanford University.

T. Bielecki – M. Rutkowski (2002), "Credit Risk: Modeling, Valuation and Hedging", Springer Verlag, New York.

F. Black – J. Cox (1976), *Valuing Corporate Securities: Liabilities: Some Effects of Bond Indenture Provisions*, Journal of Finance **31**, 351-367.

F. Black – M. Scholes (1973), *The Pricing of Options and Corporate Liabilities*, Journal of Political Economy **81**, 637-654.

Bluhm, C. – L. Overbeck – C. Wagner (2003), "An Introduction to Credit Risk Modeling", Chapman and Hall, London.

P. Brémaud (1981), "Point Processes and Queues: Martingale Dynamics", Springer, New York.

G. Chacko – L. Viceira (2003), "Spectral GMM Estimation of Continuous-Time Processes", Journal of Econometrics 116, 259-292.

R.-R. Chen – L. Scott (1995), *Interest Rate Options in Multifactor Cox-Ingersoll-Ross Models of the Term Structure*, Journal of Derivatives **3**, 53-72.

R.-R. Chen – B. Sopranzetti (1999), "The Valuation of Default-Triggered Credit Derivatives", Working Paper, Rutgers Business School, Department of Finance and Economics.

K. Chung (1982), "Lectures from Markov Processes to Brownian Motion", Springer-Verlag, New York.

P. Collin-Dufresne – R. Goldstein (2001), *Do Credit Spreads Reflect Stationary Leverage Ratios*, The Journal of Finance **56**, 1929-1957.

P. Collin-Dufresne – R. Goldstein – J. Helwege (2002), "Are Jumps in Corporate Bond Yields Priced? Modeling Contagion Via the Updating of Beliefs", Working Paper, Carnegie Mellon University.

P. Collin-Dufresne – R. Goldstein – J. Hugonnier (2004), "A General Formula for Pricing Defaultable Claims", Econometrica 72, 1347-1408.

I. COOPER – M. MARTIN (1996), *Default Risk and Derivative Products*, Applied Mathematical Finance **3**, 53-74.

I. COOPER – A. MELLO (1991), *The Default Risk of Swaps*, Journal of Finance **XLVI**, 597-620.

I. COOPER – A. MELLO (1992), "Pricing and Optimal Use of Forward Contracts with Default Risk", Working Paper, Department of Finance, London Business School, University of London.

J. COX – J. INGERSOLL – S. ROSS (1985), *A Theory of the Term Structure of Interest Rates*, Econometrica **53**, 385-408.

Q. DAI – K. SINGLETON (2000), *Specification Analysis of Affine Term Structure Models*, Journal of Finance **55**, 1943-1978.

D. DALEY – D. VERE-JONES (1988), "An Introduction to the Theory of Point Processes", Springer-Verlag, New York.

S. DAS (1998), "Poisson-Gaussian Processes and the Bond Markets", Working Paper, Department of Finance, Harvard Business School.

S. DAS – D. DUFFIE – N. KAPADIA (2004), "Common Failings: How Corporate Defaults are Correlated", Working Paper, Graduate School of Business, Stanford University.

S. DAS – R. SUNDARAM (1999), *Of Smiles and Smirks: A Term Structure Perspective*, Journal of Financial and Quantitative Analysis **34**, 211-239.

S. DAS – P. TUFANO (1995), *Pricing Credit-Sensitive Debt when Interest Rates, Credit Ratings and Credit Spreads are Stochastic"*, Journal of Financial Engineering **5** (2), 161-198.

M. DAVIS – F. LISCHKA (1999), "Convertible Bonds with Market Risk and Credit Risk", Working Paper, Tokyo-Mitsubishi International plc.

M. DAVIS – V. LO (1999), "Infectious Defaults", Working Paper, Tokyo-Mitsubishi International plc.

M. DAVIS – V. LO (2000), "Modelling Default Correlation in Bond Portfolios", Working Paper, Tokyo-Mitsubishi International plc.

M. DAVIS – T. MAVROIDIS (1997), "Valuation and Potential Exposure of Default Swaps", Working Paper, Research and Product Development, Tokyo-Mitsubishi International plc.

D. DAVYDOV – V. LINETSKY – C. LOTZ (1999), "The Hazard-Rate Approach to Pricing Risky Debt: Two Analytically Tractable Examples", Working Paper, Department of Economics, University of Michigan.

J.-P. DÉCAMPS – J.-C. ROCHET (1997), *A Variational Approach for Pricing Options and Corporate Bonds*, Economic Theory **9**, 557-569.

F. DELBAEN – W. SCHACHERMAYER (1999), *A General Version of the Fundamental Theorem of Asset Pricing*, Mathematische Annalen **300**, 463-520.

C. DELLACHERIE – P.-A. MEYER (1978), "Probabilities and Potential", North-Holland, Amsterdam.

G. DUFFEE (1999), *Estimating the Price of Default Risk*, Review of Financial Studies **12**, 197-226.

D. DUFFIE (1998a), "Defaultable Term Structures with Fractional Recovery of Par", Working Paper, Graduate School of Business, Stanford University.

D. DUFFIE (1998b), "First to Default Valuation", Working Paper, Graduate School of Business, Stanford University.

D. DUFFIE (2001), "Dynamic Asset Pricing Theory, Third Edition", Princeton University Press.

D. DUFFIE – D. FILIPOVIĆ – W. SCHACHERMAYER (2003), *Affine Processes and Applications in Finance*, Annals of Applied Probability **13**, 984-1053.

D. DUFFIE – N. GÂRLEANU (2001), *Risk and Valuation of Collateralized Debt Valuation*, Financial Analysts Journal **57** (1) January-February, 41–62.

D. DUFFIE – M. HUANG (1996), *Swap Rates and Credit Quality*, Journal of Finance **51**, 921-949.

D. DUFFIE – R. KAN (1996), *A Yield-Factor Model of Interest Rates*, Mathematical Finance **6**, 379-406.

D. DUFFIE – D. LANDO (2001), *Term Structures of Credit Spreads with Incomplete Accounting Information*, Econometrica **69**, 633-664.

D. DUFFIE – J. PAN – K. SINGLETON (2000), *Transform Analysis and Asset Pricing for Affine Jump Diffusions*, Econometrica **68**, 1343–1376.

D. DUFFIE – L. PEDERSEN – K. SINGLETON (2003), *Modeling Sovereign Yield Spreads: A Case Study of Russian Debt*, The Journal of Finance **58**, 119-160.

D. DUFFIE – M. SCHRODER – C. SKIADAS (1996), *Recursive Valuation of Defaultable Securities and the Timing of the Resolution of Uncertainty*, Annals of Applied Probability **6**, 1075-1090.

D. DUFFIE – K. SINGLETON (1999), *Modeling Term Structures of Defaultable Bonds*, Review of Financial Studies **12**, 687-720.

D. DUFFIE – K. SINGLETON (2002), "Credit Risk: Pricing, Measurement, and Management", Princeton, NJ, Princeton University Press.

D. DUFFIE – K. WANG (2003), "Multi-Period Corporate Failure Prediction with Stochastic Covariates", Working Paper, Graduate School of Business, Stanford University.

L. EDERINGTON – G. CATON – C. CAMPBELL (1997), *To Call or Not to Call Convertible Debt*, Financial Management **26**, 22-31.

R. ELLIOTT – M. JEANBLANC – M. YOR (2000), "Some Models on Default Risk", Mathematical Finance 10, 179-196.

J. ERICSSON – O. RENAULT (1999), "Credit and Liquidity Risk", Working Paper, Faculty of Management, McGill University.

H. FAN – S. SUNDARESAN (2000), *Debt Valuation, Strategic Debt Service and Optimal Dividend Policy*, Review of Financial Studies **13**, 1057-1099.

W. FELLER (1951), *Two Singular Diffusion Problems*, Annals of Mathematics **54**, 173-182.

D. FILIPOVIĆ (2001), "Time-Inhomogeneous Affine Processes", University of Munich, forthcoming in *Stochastic Processes and Their Applications*.

C. FINGER (2000), "A Comparison of Stochastic Default Rate Models", Working Paper, The Risk Metrics Group.

E. FISHER – R. HEINKEL – J. ZECHNER (1989), *Dynamic Capital Strucutre Choice: Theory and Tests*, Journal of Finance **44**, 19-40.

R. GESKE (1977), *The Valuation of Corporate Liabilities as Compound Options*, Journal of Financial Economics **7**, 63-81.

R. GIBSON – S. SUNDARESAN (1999), "A Model of Sovereign Borrowing and Sovereign Yield Spreads", Working Paper, School of HEC, University of Lausanne.

J. GRANDELL (1976), "Doubly Stochastic Poisson Processes", Lecture Notes in Mathematics, Number 529, Springer-Verlag, New York.

J. GREGORY – J. LAURENT (2003), "Basket Default Swaps, CDO's, and Factor Copulas", Working Paper, BNP Paribas.

M. HARRISON – D. KREPS (1979), *Martingales and Arbitrage in Multiperiod Securities Markets*, Journal of Economic Theory **20**, 381-408.

S. HESTON (1993), *A Closed-Form Solution for Options with Stochastic Volatility with Applications to Bond and Currency Options*, Review of Financial Studies **6**, 327-344.

J. HULL – A. WHITE (1992), *The Price of Default*, Risk **5**, 101-103.

J. HULL – A. WHITE (1995), *The Impact of Default Risk on the Prices of Options and Other Derivative Securities*, Journal of Banking and Finance **19**, 299-322.

J. JACOD – A. SHIRYAEV (1987), "Limit Theorems for Stochastic Processes", Springer-Verlag, New York.

R. JARROW – D. LANDO – S. TURNBULL (1997), *A Markov Model for the Term Structure of Credit Risk Spreads*, Review of Financial Studies **10**, 481-523.

R. JARROW – D. LANDO – F. YU (1999), "Default Risk and Diversification: Theory and Empirical Applications", Cornell University, forthcoming *in Mathematical Finance*.

R. JARROW – S. TURNBULL (1995), *Pricing Derivatives on Financial Securities Subject to Credit Risk*, Journal of Finance **50**, 53-85.

R. JARROW – F. YU (2001), *Counterparty Risk and the Pricing of Defaultable Securities*, Journal of Finance **56**, 1765-1800.

M. JEANBLANC – M. RUTKOWSKI (2000), "Modelling of Default Risk: An Overview", Modern Mathematical Finance: Theory and Practice. Higher Education Press, Beijing, 171-269.

G. Jiang – J. Knight (2002), "Estimation of Continuous Time Processes Via Empirical Characteristic Function", Journal of Business and Economic Statistics 20, 198-212.

I. Karatzas – S. Shreve (1988), "Brownian Motion and Stochastic Calculus", Springer-Verlag, New York.

A. Karr (1991), "Point Processes and Their Statistical Inference", 2d ed., Marcel Dekker, Inc, New York.

K. Kawazu – S. Watanabe (1971), Branching Processes with Immigration and Related Limit Theorems", Theory Probab. Appl. **16**, 36-54.

M. Kijima (1998), Monotonicities in a Markov Chain Model for Valuing Corporate Bonds Subject to Credit Risk, Mathematical Finance **8**, 229-247.

M. Kijima – K. Komoribayashi (1998), A Markov Chain Model for Valuing Credit Risk Derivatives, Journal of Derivatives **6** (Fall), 97-108.

S. Kusuoka (1999), A Remark on Default Risk Models, Advances in Mathematical Economics **1**, 69-82.

D. Lando (1994), "Three Essays on Contingent Claims Pricing", Working Paper, Ph.D. Dissertation, Statistics Center, Cornell University.

D. Lando (1998), On Cox Processes and Credit Risky Securities, Review of Derivatives Research **2**, 99-120.

D. Lando (2004), "Credit Risk Modeling", Princeton, NJ, Princeton University Press.

D. Lando – T. Skødeberg (2002), "Analyzing Rating Transitions and Rating Drift with Continuous Observations", Journal of Banking and Finance 26, 423-444.

H. Leland (1994), Corporate Debt Value, Bond Covenants, and Optimal Capital Structure, Journal of Finance **49**, 1213-1252.

H. Leland (1998), Agency Costs, Risk Management, and Capital Structure, Journal of Financ **53**, 1213-1242.

H. Leland – K. Toft (1996), Optimal Capital Structure, Endogenous Bankruptcy, and the Term Structure of Credit Spreads, Journal of Finance **51**, 987-1019.

R. Litterman – T. Iben (1991), Corporate Bond Valuation and the Term Structure of Credit Spreads, Journal of Portfolio Management Spring, 52-64.

F. Longstaff – E. Schwartz (1995a), A Simple Approach to Valuing Risky Fixed and Floating Rate Debt, Journal of Finance **50**, 789-819.

F. Longstaff – E. Schwartz (1995b), Valuing Credit Derivatives, Journal of Fixed Income **5** (June), 6-12.

B. Loshak (1996), "The Valuation of Defaultable Convertible Bonds under Stochastic Interest Rate", Working Paper, Krannert Graduate School of Management, Purdue University, West Lafayette.

D. Madan – H. Unal (1998), Pricing the Risks of Default, Review of Derivatives Research **2**, 121-160.

G. MANSO – B. STRULOVICI – A. TCHISTYI (2003), "Performance Sensitive Debt and the Credit Cliff Dynamic", Working Paper, Graduate School of Business, Stanford University.

P. MELLA-BARRAL (1999), *Dynamics of Default and Debt Reorganization*, Review of Financial Studies **12**, 535-578.

P. MELLA-BARRAL – W. PERRAUDIN (1997), *Strategic Debt Service*, Journal of Finance **52**, 531-556.

J. MERRICK (1999), "Crisis Dynamics of Russian Eurobond Implied Default Recovery Ratios", Working Paper, Stern School of Business, New York University.

R. MERTON (1974), *On the Pricing of Corporate Debt: The Risk Structure of Interest Rates"*, Journal of Finance **29**, 449-470.

P.-A. MEYER (1966), "Probability and Potentials", Waltham, MA: Blaisdell Publishing Company.

F. MODIGLIANI – M. MILLER (1958), *The Cost of Capital, Corporation Finance, and the Theory of Investment*, American Economic Review **48**, 261-297.

L. NIELSEN – J. SAÁ-REQUEJO – P. SANTA-CLARA (1993), "Default Risk and Interest Rate Risk: The Term Structure of Credit Spreads", Working Paper, INSEAD, Fontainebleau, France.

S. NIELSEN – E. RONN (1995), "The Valuation of Default Risk in Corporate Bonds and Interest Rate Swaps", Working Paper, Department of Management Science and Information Systems, University of Texas at Austin.

K. NYBORG (1996), *The Use and Pricing of Convertible Bonds*, Applied Mathematical Finance **3**, 167-190.

H. PAGÈS (2000), "Estimating Brazilian Sovereign Risk from Brady Bond Prices", Working Paper, Bank of France.

J. PAN (2002), *The Jump-Risk Premia Implicit in Options: Evidence from an Integrated Time-Series Study*, Journal of Financial Economics **63**, 3-50.

Y. PIERIDES (1997), *The Pricing of Credit Risk Derivatives*, Journal of Economic Dynamics and Control **21**, 1579-1611.

P. PROTTER (2004), "Stochastic Integration and Differential Equations, Second Edition", Springer-Verlag, New York.

G. PYE (1974), *Gauging the Default Premium*, Financial Analysts Journal (January-February), 49-52.

D. REVUZ – M. YOR (1994), "Continuous Martingales and Brownian Motion", Volume 293 of "Grundlehren der mathematischen Wissenschaften", Springer-Verlag, Berlin-Heidelberg-New York.

K. SATO (1999), "Lévy processes and infinitely divisible distributions", Cambridge University Press, Cambridge. Translated from the 1990 Japanese original, Revised by the author.

P. SCHÖNBUCHER (1998), *Term Stucture Modelling of Defaultable Bonds*, Review of Derivatives Research **2**, 161-192.

P. Schönbucher (2003a), "Credit Derivatives and Pricing Models", Wiley, New York.

P. Schönbucher – D. Schubert (2001), "Copula-Dependent Default Risk in Intensity Models", Working Paper, Department of Statistics, Bonn University.

P. J. Schönbucher (2003b), "Information-Driven Default Contagion", Working Paper, Department of Mathematics, ETH Zurich.

L. Scott (1997), *Pricing Stock Options in a Jump-Diffusion Model with Stochastic Volatility and Interest Rates: Applications of Fourier Inversion Methods*, Mathematical Finance **7**, 413-426.

T. Shiga – S. Watanabe (1973), *Bessel Diffusions as a One-Parameter Family of Diffusion Processes*, Z. Wahrscheinlichkeitstheorie verw. Geb. **27**, 37-46.

D. Shimko – N. Tejima – D. van Deventer (1993), *The Pricing of Risky Debt when Interest Rates are Stochastic*, Journal of Fixed Income **3** (September), 58–65.

K. Singleton (2001), *Estimation of Affine Asset Pricing Models using the Empirical Characteristic Function*, Journal of Econometrics **102**, 111-141.

S. Song (1998), "A Remark on a Result of Duffie and Lando", Working Paper, Department of Mathematics, Université d'Evry, France.

E. Stein – J. Stein (1991), *Stock Price Distributions with Stochastic Volatility: An Analytic Approach*, Review of Financial Studies **4**, 725-752.

K. Tsiveriotis – C. Fernandes (1998), *Valuing Convertible Bonds With Credit Risk*, Journal of Fixed Income **8**, 95-102.

M. Uhrig-Homburg (1998), "Endogenous Bankruptcy when Issuance is Costly", Working Paper, Department of Finance, University of Mannheim.

O. Vasicek (1977), *An Equilibrium Characterization of the Term Structure*, Journal of Financial Economics **5**, 177-188.

S. Watanabe (1969), *On two dimensional Markov processes with branching property*, Trans. Amer. Math. Soc. **136**, 447-466.

F. Yu (2002a), "Accounting Transparency and the Term Structure of Credit Spreads", University of California, Irvine, forthcoming *Journal of Financial Economics*.

F. Yu (2002b), "Default Correlation in Reduced-Form Models", Working Paper, University of California, Irvine.

C.-S. Zhou (2001), "The Term Structure of Credit Spreads with Jump Risk", Journal of Banking and Finance 25, 2015-2040.

PUBBLICAZIONI DELLA CLASSE DI SCIENZE
DELLA SCUOLA NORMALE SUPERIORE
QUADERNI

1. DE GIORGI E., COLOMBINI F., PICCININI L.C.: *Frontiere orientate di misura minima e questioni collegate.*
2. MIRANDA C.: *Su alcuni problemi di geometria differenziale in grande per gli ovaloidi.*
3. PRODI G., AMBROSETTI A.: *Analisi non lineare.*
4. MIRANDA C.: *Problemi in analisi funzionale* (ristampa).
5. TODOROV I.T., MINTCHEV M., PETKOVA V.B.: *Conformal Invariance in Quantum Field Theory.*
6. ANDREOTTI A., NACINOVICH M.: *Analytic Convexity and the Principle of Phragmén-Lindelöf.*
7. CAMPANATO S.: *Sistemi ellittici in forma divergenza. Regolarità all'interno.*
8. *TOPICS IN FUNCTIONAL ANALYSIS:* Contributors: F. STROCCHI, E. ZARANTONELLO, E. DE GIORGI, G. DAL MASO, L. MODICA.
9. LETTA G.: *Martingales et intégration stochastique.*
10. *OLD AND NEW PROBLEMS IN FUNDAMENTAL PHYSICS:* Meeting in honour of GIAN CARLO WICK.
11. *INTERACTION OF RADIATION WITH MATTER:* A Volume in honour of ADRIANO GOZZINI.
12. MÉTIVIER M.: *Stochastic Partial Differential Equations in Infinite Dimensional Spaces.*
13. *SYMMETRY IN NATURE:* A Volume in honour of LUIGI A. RADICATI DI BROZOLO.
14. *NONLINEAR ANALYSIS:* A Tribute in honour of GIOVANNI PRODI.
15. LAURENT-THIÉBAUT C., LEITERER J.: *Andreotti-Grauert Theory on Real Hypersurfaces.*
16. ZABCZYK J.: *Chance and Decision. Stochastic Control in Discrete Time.*
17. EKELAND I.: *Exterior Differential Calculus and Applications to Economic Theory.*
18. *ELECTRONS AND PHOTONS IN SOLIDS:* A Volume in honour of FRANCO BASSANI.
19. ZABCZYK J.: *Topics in Stochastic Processes.*
20. TOUZI N.: *Stochastic Control Problems, Viscosity Solutions and Application to Finance.*

CATTEDRA GALILEIANA

1. LIONS P.L.: *On Euler Equations and Statistical Physics.*
2. BJÖRK T.: *A Geometric View of the Term Structure of Interest Rates.*
3. DELBAEN F.: *Coherent Risk Measures.*
4. SCHACHERMAYER W.: *Portfolio Optimization in Incomplete Financial Markets.*
5. DUFFIE D.: *Credit Risk Modeling with Affine Processes.*

LEZIONI LAGRANGE

1. VOISIN C.: *Variations of Hodge Structure of Calabi-Yau Threefolds.*

LEZIONI FERMIANE

1. THOM R.: *Modèles mathématiques de la morphogénèse.*
2. AGMON S.: *Spectral Properties of Schrödinger Operators and Scattering Theory.*
3. ATIYAH M.F.: *Geometry of Yang-Mills Fields.*
4. KAC M.: *Integration in Function Spaces and Some of Its Applications.*
5. MOSER J.: *Integrable Hamiltonian Systems and Spectral Theory.*
6. KATO T.: *Abstract Differential Equations and Nonlinear Mixed Problems.*
7. FLEMING W.H.: *Controlled Markov Processes and Viscosity Solution of Nonlinear Evolution Equations.*

8. ARNOLD V.I.: *The Theory of Singularities and Its Applications.*
9. OSTRIKER J.P.: *Development of Larger-Scale Structure in the Universe.*
10. NOVIKOV S.P.: *Solitons and Geometry.*
11. CAFFARELLI L.A.: *The Obstacle Problem.*
12. CHEEGER J.: *Degeneration of Riemannian metrics under Ricci curvature bounds.*

PUBBLICAZIONI DEL CENTRO DI RICERCA MATEMATICA ENNIO DE GIORGI

1. DYNAMICAL SYSTEMS. Part I: *Hamiltonian Systems and Celestial Mechanics.*
2. DYNAMICAL SYSTEMS. Part II: *Topological, Geometrical and Ergodic Properties of Dynamics.*
3. MATEMATICA, CULTURA E SOCIETÀ 2003.
4. RICORDANDO FRANCO CONTI.
5. N. KRYLOV: *Probabilistic Methods of Investigating Interior Smoothness of Harmonic Functions Associated with Degenerate Elliptic Operators.*

ALTRE PUBBLICAZIONI

Proceedings of the Symposium on FRONTIER PROBLEMS IN HIGH ENERGY PHYSICS Pisa, June 1976.
Proceedings of International Conferences on SEVERAL COMPLEX VARIABLES, Cortona, June 1976 and July 1977.
Raccolta degli scritti dedicati a JEAN LERAY apparsi sugli Annali della Scuola Normale Superiore di Pisa.
Raccolta degli scritti dedicati a HANS LEWY apparsi sugli Annali della Scuola Normale Superiore di Pisa.
Indice degli articoli apparsi nelle Serie I, II e III degli Annali della Scuola Normale Superiore di Pisa (dal 1871 al 1973).
Indice degli articoli apparsi nella Serie IV degli Annali della Scuola Normale Superiore di Pisa (dal 1974 al 1990).
ANDREOTTI A.: *SELECTA vol. I, Geometria algebrica..*
ANDREOTTI A.: *SELECTA vol. II, Analisi complessa, Tomo I e II..*
ANDREOTTI A.: *SELECTA vol. III, Complessi di operatori differenziali..*

Fotocomposizione "CompoMat" Loc. Braccone, 02040 Configni (RI), Italy
Finito di stampare per conto della "CompoMat" dalla Nuova Grafica 86 nel dicembre 2004